Concepts and Applications of Thermal Engineering

Concepts and Applications of Thermal Engineering

Felix Dom

www.willfordpress.com

Published by Willford Press,
118-35 Queens Blvd., Suite 400,
Forest Hills, NY 11375, USA

ISBN: 978-1-64728-505-0

Cataloging-in-Publication Data

Concepts and applications of thermal engineering / Felix Dom.
p. cm.
Includes bibliographical references and index.
ISBN 978-1-64728-505-0
1. Heat engineering. 2. Thermodynamics. 3. Power (Mechanics). I. Dom, Felix.
TJ260 .T44 2023
621.402 2--dc23

For information on all Willford Press publications
visit our website at www.willfordpress.com

Contents

Preface

Over the recent decade, advancements and applications have progressed exponentially. This has led to the increased interest in this field and projects are being conducted to enhance knowledge. The main objective of this book is to present some of the critical challenges and provide insights into possible solutions. This book will answer the varied questions that arise in the field and also provide an increased scope for furthering studies.

Thermal engineering refers to a sub-discipline of mechanical engineering concerned with the transmission and movement of heat energy. This energy can be converted into different types of energy. Thermal engineering is used in a variety of industries including the construction industry, heating ventilation and cooling industry, and automotive industry. Concepts related to mass transfer, heat transfer, thermodynamics and fluid mechanics may be used to solve thermal engineering problems. There are various applications of thermal engineering including process fired heaters, thermal power plants, combustion engines, compressed air systems, thermal insulation, boiler design, refrigeration systems, and heat exchangers. This book outlines the concepts and applications of thermal engineering in detail. It is a valuable compilation of topics, ranging from the basic to the most complex advancements in this field. This book will serve as a reference to a broad spectrum of readers.

I hope that this book, with its visionary approach, will be a valuable addition and will promote interest among readers. Each of the authors has provided their extraordinary competence in their specific fields by providing different perspectives as they come from diverse nations and regions. I thank them for their contributions.

Felix Dom

1

Review of First and Second Laws

1.1 First Law Analysis of Unsteady Flow Control Volumes

For any system and in any process, the first law can be written as,

$$Q = \Delta E + W$$

Where,

E -Represents all forms of energy stored in the system.

For a pure substance,

$$E = E_K + E_p + U$$

Where,

E_K the K.E., E_p the P.E. and U the residual energy stored in the molecular structure of the substance.

$$Q = \Delta E_K + \Delta E_P + \Delta U + W \quad \text{...(i)}$$

When there is mass transfer across the system boundary, the system is called an open system. Most of the engineering devices are open systems involving the flow of fluids through them.

Equation (i) refers to a system having a particular mass of substance and it is free to move from place to place.

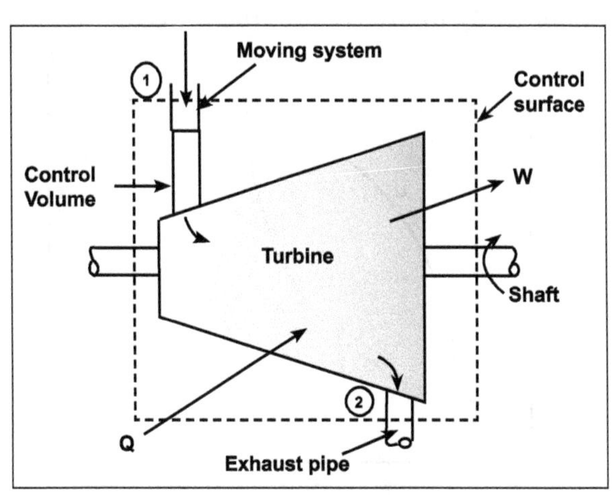

Flow process involving work and heat interaction.

Consider a steam turbine shown in the figure, in which steam enters at a high pressure, does work upon the turbine rotor and then leaves the turbine at low pressure through the exhaust pipe. If a certain mass of steam is considered as the thermodynamic system, then the energy equation becomes,

$$Q = \Delta E_K + \Delta E_P + \Delta U + W$$

and in order to analyse the expansion process in turbine the moving system is to be followed as it travels through the turbine, taking into account the work and heat interactions all the way through. This method of analysis is similar to that of Lagrange in fluid mechanics.

Although the system approach is quite valid, there is another approach which is found to be highly convenient. Instead of concentrating attention upon a certain quantity of fluid, which constitutes a moving system in flow processes, attention is focused upon a certain fixed region in space called a control volume through which the moving substance flows. This is similar to the analysis of Euler in fluid mechanics.

To distinguish the two concepts, it may be noted that while the system (closed) boundary usually changes shape, position and orientation relative to the observer, the control volume boundary remains fixed and unaltered. Again, while matter usually crosses the control volume boundary, no such flow occurs across the system boundary.

The broken line in the figure represents the surface of the control volume which is known as the control surface. This is the same as the system boundary of the open system. The method of analysis is to inspect the control surface and account for all energy quantities transferred through this surface. Since there is mass transfer across the control surface, a mass balance also has to be made.

1.1.1 Entropy Change for Different Process and Entropy Generation

We know that change in entropy in a reversible process is equal to equation given below,

$$\left(\frac{\delta Q}{T}\right)_R \quad \text{...(i)}$$

Let us now find the change in entropy in an irreversible process.

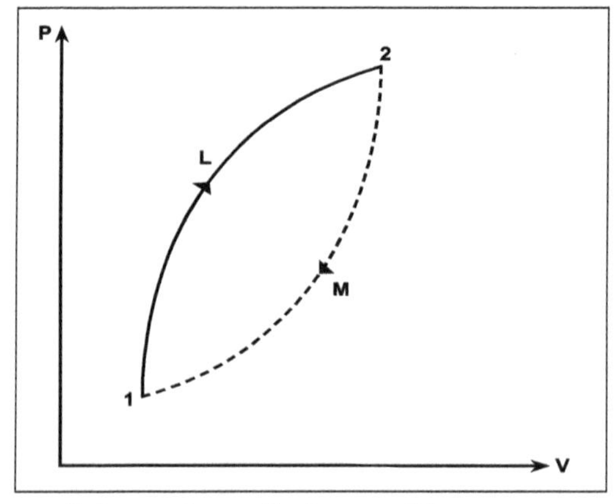

Entropy change for irreversible process.

Consider a closed system undergoing a change from state 1 to state 2 by a reversible process 1-L-2 and returns from state 2 to the initial state 1 by an irreversible process 2-M-1 as shown in the figure, above on the thermodynamic coordinates, pressure and volume.

Since entropy is a thermodynamic property, we can write,

$$\oint dS = \int_{1(L)}^{2} (dS)_R + \int_{2(M)}^{1} (dS)_I = 0 \quad \text{...(ii)}$$

(Subscript I represent the irreversible process).

Now for a reversible process, from equation (i), we have,

$$\int_{1(L)}^{2} (dS)_R = \int_{1(L)}^{2} \left(\frac{\delta Q}{T}\right)_R \quad \text{...(iii)}$$

Substituting the value of $\int_{1(L)}^{2} (dS)_R$ in equation (ii), we get

$$\int_{1(L)}^{2} \left(\frac{\delta Q}{T}\right)_R + \int_{2(M)}^{1} (dS)_I = 0 \quad \text{...(iv)}$$

Again, since in equation (ii) the processes 1-L-2 and 2-M-1 together form an irreversible cycle, applying Clausius equality to this expression, we get

$$\oint \frac{\delta Q}{T} = \int_{1(L)}^{2} \left(\frac{\delta Q}{T}\right)_R + \int_{2(M)}^{1} \left(\frac{\delta Q}{T}\right)_I < 0 \quad \text{...(v)}$$

Now subtracting equation (v) from equation (iv), we get

$$\int_{2(M)}^{1} (dS)_I > \int_{2(M)}^{1} \left(\frac{\delta Q}{T}\right)_I$$

which for infinitesimal changes in states can be written as,

$$(dS)_I > \int \left(\frac{\delta Q}{T}\right)_I \quad \text{...(vi)}$$

Equation (vi) states that the change in entropy in an irreversible process is greater than $\delta Q / T$ Combining equation (v) and (vi), we can write the equation in the general form as,

$$(dS) > \left(\frac{\delta Q}{T}\right) \quad \text{...(vii)}$$

Where, equality sign stands for the reversible process and inequality sign stands for the irreversible process.

It may be noted here that the effect of irreversibility is always to increase the entropy of the system. Let us now consider an isolated system. We know that in an isolated system, matter, work or heat cannot cross the boundary of the system. Hence according to first law of thermodynamics, the internal energy of the system will remain constant.

Since for an isolated system, $\delta Q = 0$, from equation(vii), we get:

$$(dS)_{isolated} \geq 0 \text{ (viii)}$$

Equation (viii) states that the entropy of an isolated system either increases or remains constant. This is a corollary of the second law. It explains the principle of increase in entropy.

Entropy Generation

Consider a system undergoing a process 1-2 as shown in the diagram. The increase of entropy principle states:

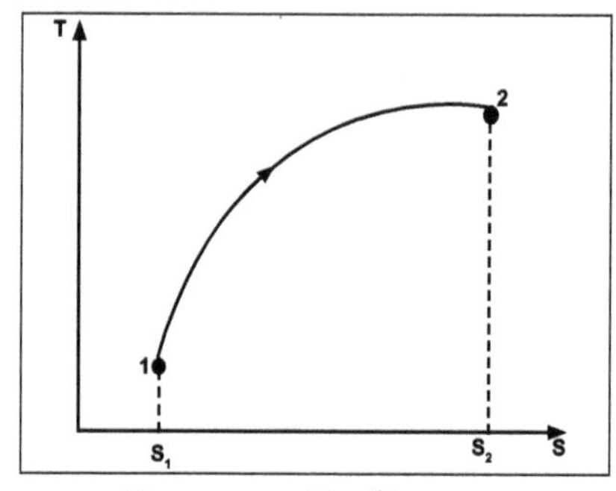

Process 1-2 T-s diagram.

$$\Delta S = S_2 - S_1 \geq \int_1^2 \frac{\delta Q}{T}$$

The above equation can be expressed as,

$$\Delta S = S_2 - S_1 = \int_1^2 \frac{\delta Q}{T} + S_{gen}$$

Where,

S_{gen} - Entropy generation

S_{gen} measures the effect of irreversibilities present within the system during a process. Its value depends on the nature of the process and not solely on the start and end states. Hence, it is not a property. Rewriting the above equation gives the expression of entropy generation as,

$$S_{gen} = S_2 - S_1 - \int_1^2 \frac{\delta Q}{T}$$

Non-negative Entropy Generation,

$S_{gen} > 0$ Irreversible process,

$S_{gen} = 0$ internally reversible process,

$S_{gen} < 0$ Impossible

According to the increase of entropy principle, entropy generation can only take a non-negative value.

1.1.2 Entropy Balance for Closed Systems and Steady Flow Systems

Entropy Balance for Closed Systems

A closed system does not involve mass transfer across its boundaries thus its entropy change during a process is the sum of entropy transfer due to heat transfer and entropy generation within a system. Thus,

$$S_2 - S_1 = \int \frac{\delta Q}{T} + S_{gen}$$

$$S_2 - S_1 = \sum \frac{Q}{T} + S_{gen}$$

For an adiabatic process (Q = 0), the above expression reduces to,

$$S_2 - S_1 = S_{gen}$$
$$(\Delta S)_{adiabatic} = S_{gen}$$

Entropy Change for Open Systems

An open system involves mass transfer and heat transfer across its boundaries. Therefore, the rate of entropy change of an open system is the sum of net rate of entropy transport due to mass flow, entropy-transfer rate due to heat transfer at its boundaries and the net rate of entropy generation within the system due to irreversibilities involved in the process.

Thus, for one-dimensional flow,

$$\frac{dS}{dt} = \sum \dot{m}_i s_i - \sum \dot{m}_e s_e + \sum \frac{\dot{Q}}{T} + \dot{S}_{gen}$$

For a steady flow device, $\frac{dS}{dt} = 0$

$$\dot{S}_{gen} = \dot{m}(s_e - s_i) - \sum \frac{\dot{Q}}{T}$$

Steady Flow System

As a fluid flows through a certain control volume, its thermodynamic properties may vary along the space coordinates as well as with time. If the rates of flow of mass and energy through the control surface change with time, the mass and energy within the control volume also would change with time. 'Steady flow' means that the rates of flow of mass and energy across the control surface are constant. In most engineering devices, they are a constant rate of flow of mass and energy through the control surface and the control volume in course of time attains a steady state.

At the steady state of a system, any thermodynamic property will have a fixed value at a particular location and will not alter with time. Thermodynamic properties may vary along space coordinates, but do not vary with time. 'Steady state' means that the state is steady or invariant with time.

1.2 Available Energy and Quality of Energy

Available Energy

From the second law of thermodynamics, the heat-energy supplied to a system cannot be fully converted to work energy. The maximum work which can be obtained from the heat supplied is called the available energy. The rest of the heat energy which cannot be converted to work is called unavailable energy. Josiah Willard Gibbs is the originator of the availability concept. He indicated that environment plays an important role in evaluating the available energy. The available energy is also known as Ex-energy and the unavailable energy.

Quality of Energy

Let us assume that a hot gas flows through a pipeline shown in the figure (a). Due to heat loss to the surroundings, the temperature of the gas decreases, continuously from inlet at state A to the exit at state B. Let us assume a reversible isobaric path between the inlet and exit states of the gas an shown in the figure (b).

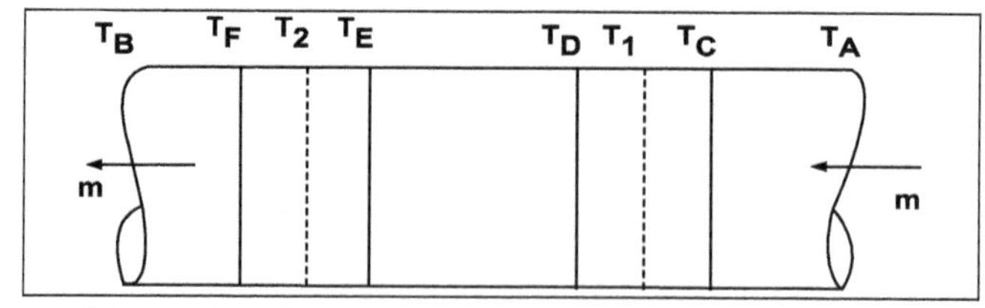

(a) Heat loss from a hot gas flowing through a pipeline.

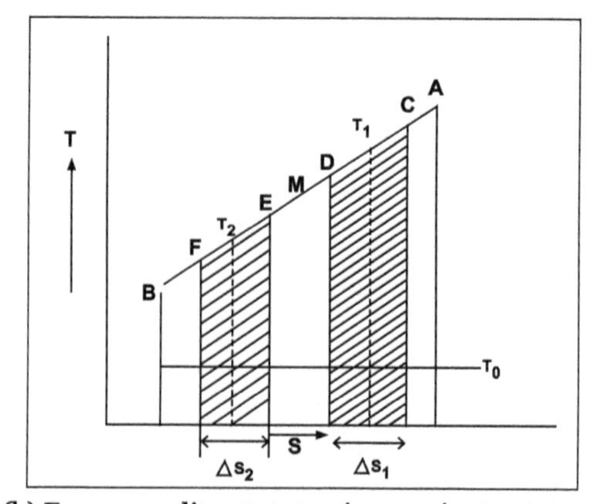

(b) Energy quality at state 1 is superior to state 2.

For an infinitesimal reversible process at constant pressure,

$$dS = \frac{dQ}{T} = \frac{mC_p dT}{T}$$

$$\frac{dT}{S} = \frac{T}{mC_p} \qquad \text{...(i)}$$

Let us consider that Q amounts of heat are lost to the surroundings as the temperature of the gas decreases from T_C to T_D.

Let T_1 be the average of T_C and T_D. Then heat loss,

$$Q = mC_p (T_c - T_D) = T_1 \Delta S_1 \qquad \text{...(ii)}$$

Available energy lost with this heat, i.e., heat lost at temperature T1 is given by,

$$W_1 = Q - T_0 \Delta S_1 \qquad \text{...(iii)}$$

When, the gas temperature has reached T_2($T_2 < T_1$). Let us assume that the same heat loss Q occurs as the gas temperature decreases from T_E to T_F Let T_2 be the average temperature of T_E and T_F Then,

Heat loss,

$$Q = mC_p (T_E - T_F)$$
$$= T_2 \Delta S_2 \qquad \text{...(iv)}$$

Available energy lost with this heat loss at temperature T2 is given by,

$$W_2 = Q - T_0 \Delta S_2 \qquad \text{...(v)}$$

Now from (ii) and (iv)

$$T_1 \Delta S_1 = T_2 \Delta S_2, \text{but } T_1 > T_2$$
$$\Delta S_1 < \Delta S_2$$

Hence, using (iii) and (v)

$$\Delta S_1 < \Delta S_2$$

$$W_1 > W_2 \qquad \text{...(vi)}$$

From equation (vi), it is concluded that loss of available energy in more when heat loss occurs at a higher temperature T_1 than at temperature T_2.

Thus the available energy or ex-energy of a fluid at a higher temperature T_1 is more than that at a lower temperature T_2 and decreases as the temperature decreases. The concept of available energy or ex-energy provides a useful measure of this energy quality.

1.2.1 Availability for Non-Flow and Flow Process

Availability of Steady Flow Process

In a steady flow process for a single flow referring to equation (i), we have

$$W_{max} = W_{reversible} = \left(H_1 - T_0S_1 + \frac{mV_1^2}{2} + mgz_1\right) - \left(H_2 - T_0S_2 + \frac{mV_2^2}{2} + mgz_2\right)$$

$$W_{max} = \left(H_1 - T_0S_1 + \frac{mV_1^2}{2} + mgz_1\right) - \left(H_2 - T_0S_2 + \frac{mV_2^2}{2} + mgz_2\right) \quad \text{...(i)}$$

With a given state for the mass entering the control volume, the maximum available work (availability) would be when this mass leaves the control volume in equilibrium with the surroundings, i.e., at dead state having zero velocity and minimum potential energy.

Let us assign the velocity at entry V and exit velocity, i.e., at dead state zero and z_1 as z and z_2 as z_0.

The maximum work available or availability A is given by,

$$A = \left(H - T_0S + \frac{mV^2}{2} + mgz\right) - \left(H_0 - T_0S_0 + mgz_0\right)$$

$$= \left(\psi - \psi_0\right)$$

Where,

Ψ - is called the availability function for a steady flow system.

For unit mass,

$$A = \left(h_1 - T_0S + \frac{v^2}{2} + gz\right) - \left(h_0 - T_0S_0 + gz\right)$$

Availability For Non-Flow and Flow Process

A closed system denoting the initial state as $E_1 = E, V_1 = V, Z_1 = Z$ and the final dead state $E_2 = E_0$, $V_2 = V_0 = 0$, $Z_2 = Z_0$, the equation given by,

$$\left(W_{available}\right)_{max} = E_1 - E_2 + P_0(V_1 - V_2) - T_0(S_1 - S_2)$$

$$= \left(u + \frac{mV^2}{2} + mgz\right) - \left(u_0 + mgz_0\right) + P_0\left(V - V_0\right) - \left(S - S_0\right)$$

It can be expressed as, Availability $A = \left(W_{available}\right)_{max} = E - E_0 + P_0\left(V - V_0\right) - T_0\left(S - S_0\right)$.

Neglecting KE and PE changes and considering unit mass availability,

$$A = u - u_o + P_o(V - V_o) - T_o(S - S_o)$$
$$= (u + p_o V - T_o S) - (u_o + P_o V_o - T_o S_o)$$
$$= (\phi - \phi_o)$$

Where, ϕ is called the availability function of the closed system.

1.2.2 Irreversibility and Second Law Efficiency

Reversibility and Irreversibility

The second law of thermodynamics enables us to divide all processes into two classes:

- Reversible or ideal process.
- Irreversible or natural process.

A reversible process is one which is performed in such a way that at the conclusion of the process, both the system and the surroundings may be restored to their initial states, without producing any changes in the rest of the universe. Let the state of a system be represented by, a figure given below and let the system be taken to state B quasi statically by following the path A-B.

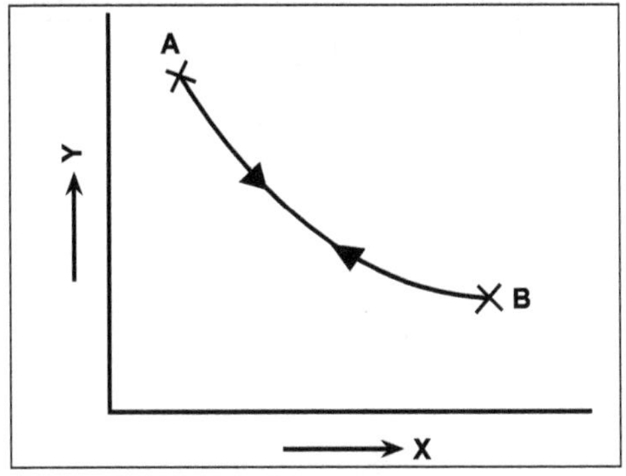

Reversible processes.

If the system and also the surroundings are restored to their initial states on reverse and no change in the universe is produced, then the process A-B will be a reversible process. In the reverse process, the system has to be taken from state B to A by following the same path B-A. A reversible process should not leave any trace or relic behind to show that the process had ever occurred.

A reversible process is carried out infinitely slowly with an infinitesimal gradient, so that every slate passed through by the system is an equilibrium state. So a reversible process coincides with a quasi-static process. Any natural process carried out with a finite gradient is an irreversible process.

A reversible process, which consists of a succession of equilibrium stales, is an idealized hypothetical process, approached only as a limit. It is said to be an asymptote to reality. All spontaneous processes are irreversible. Time has an important effect on reversibility.

If the time allowed for a process to occur is infinitely large, even though the gradient is finite, the process becomes reversible. However, if this time is squeezed to a finite value, the finite gradient makes the process irreversible.

Second Law Efficiency

The first law efficiency ηI is defined as the ratio of the output energy of a machine or device to the input energy applied to the machine or device. Thus,

$$\eta_I = \frac{\text{output energy}}{\text{input energy}}$$

The first law of thermodynamics is concerned only with the quantity of energy, whereas the second law of thermodynamics is concerned with the quality of energy. The concept of available energy provides a useful measure of the energy quality.

The second law efficiency, ηII is defined as the ratio of the minimum available energy consumed to perform any work to the actual amount of the available energy consumed in performing the work.

Thus,

$$\eta_I = \frac{\text{The minimum available energy for perform the work}}{\text{The actual energy in performing the work}}$$

$$= \frac{A_{minimum}}{A}, \text{Where A is the available energy}$$

Now, it is known that

$$\text{Irreversibility} = I = W_{max} - W \text{ (useful work) and then,}$$

$$\eta_{II} = \frac{W}{W_{max}}$$

$$\eta_{II} = \frac{W}{Q_I} = \frac{W}{W_{max}} \times \frac{W_{max}}{Q_I}$$

$$= \eta_{II} \times \eta_{Carnot}$$

$$\eta_{II} = \frac{\eta_I}{\eta_{Carnot}}$$

$$\text{Since, } W_{max} = Q_1\left(1 - \frac{T_o}{T}\right)$$

Where, To is the surrounding temperature.

$$\text{Hence, } \eta_{II} = \frac{W}{Q_1\left(1 - \frac{T_o}{T}\right)} = \frac{\eta_I}{\eta_{Carnot}}$$

Let as consider a case where the heat energy Q_r is available from a reservoir at temperature T_r and if the quantity of the heat energy Q_a in transformed at a temperature T_a then,

$$\eta_{II} = \frac{Q_a}{Q_r} \text{ and } \eta_I = \frac{\text{Available energy output}}{\text{Available energy input}}$$

$$= \frac{Q_a\left(1-\frac{T_o}{T_a}\right)}{Q_r\left(1-\frac{T_o}{T_r}\right)}$$

T_o is the surrounding temperature.

$$= \frac{Q_a}{Q_r} \times \frac{\left(1-\frac{T_o}{T_a}\right)}{\left(1-\frac{T_o}{T_r}\right)} = \eta_1 \times \frac{\left(1-\frac{T_o}{T_a}\right)}{\left(1-\frac{T_o}{T_r}\right)}$$

2

Air Standard Cycle, I.C. Engine and Reciprocating Air Compressors

2.1 Air Standard Cycle and Introduction to I.C. Engine

The assumptions made for air standard cycle analysis are:

- Air is the working fluid and it obeys the perfect gas law $PV = mRT$.
- The engine operates in a closed cycle. The cylinder is filled with constant amount of working medium and the same fluid is used repeatedly and hence mass remains constant.
- The working fluid is homogeneous throughout at all times and no chemical reaction takes place, inside the cylinder.
- The compression and expansion processes are assumed to be adiabatic.

Otto Cycle

Otto cycle is a gas power cycle that is used in SI engines (modern petrol engines). This cycle was introduced by Dr. Nikolaus August Otto, a German Engineer.

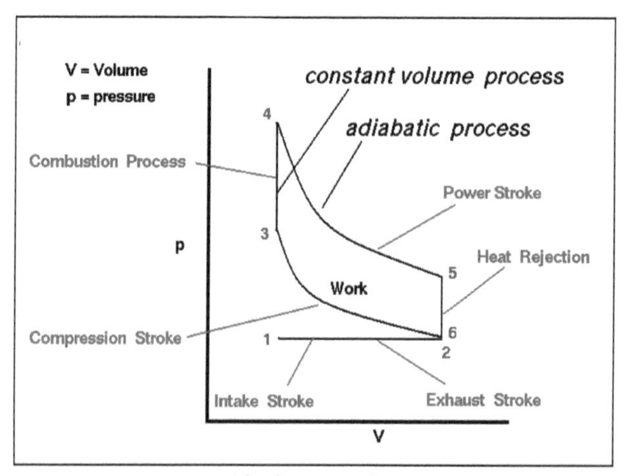

ideal otto cycle.

An Otto cycle consists of four processes:

- Two isentropic (reversible adiabatic) processes
- Two isochoric (constant volume) processes

These processes can be easily understood if we understand p-V (Pressure-Volume) and T-s (Temperature-Entropy) diagrams of Otto cycle.

The PV and the TS Diagram for the Otto Cycle

The Otto cycle is also called a constant volume or explosion cycle. This is the equivalent air cycle for reciprocating piston engines using spark ignition. Figures (a) and (b) shows the p-V and T-s diagrams respectively.

Processes in Otto Cycle

Process 1-2: Isentropic Compression

A process, the piston moves from bottom dead centre (BDC) to top dead centre (TDC) position. Air undergoes reversible adiabatic (isentropic) compression. We know that compression is a process in which volume decreases and then pressure increases.

Hence, in this process, volume of air decreases from V1 to V2 and pressure increases from p1 to p2.

Temperature increases from T_1 to T_2. As this an isentropic process, entropy remains constant (i.e., $s_1 = s_2$). Refer p-V and T-s diagrams for better understanding.

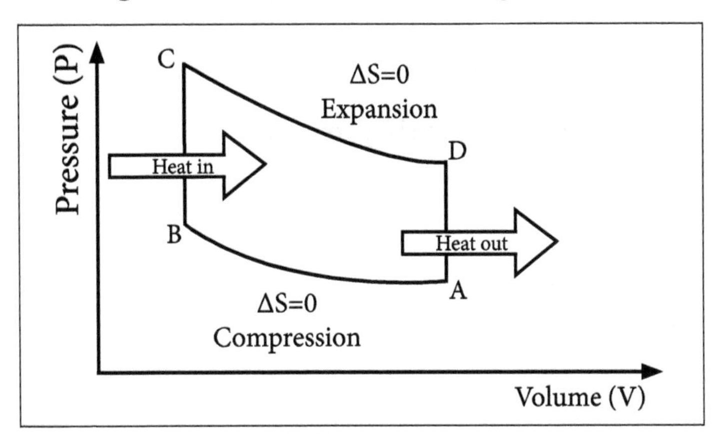

(a) PV-diagram.

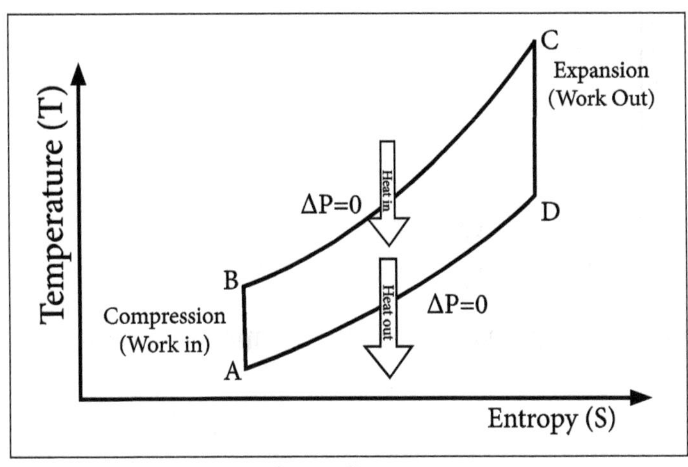

(b) TS-diagram.

Process 2-3: Constant Volume Heat Addition

In this process, isochoric (constant volume) heat addition piston remains at top dead centre for a moment. Heat is added at constant volume ($V_2 = V_3$) from an external heat source. Temperature increases from T_2 to T_3, pressure increases from p_2 to p_3 and entropy increases from S_3 to S_4.

Process 3-4: Isentropic Expansion

In this process, air undergoes isentropic (reversible adiabatic) expansion. The piston is pushed from top dead centre (TDC) to bottom dead centre (BDC) position. Here, pressure decreases p_3 to p_4, volume increases from v_3 to v_4, temperature falls from T_3 to T_4 and entropy remains constant ($s_3 = s_4$).

Process 4-1: Constant Volume Heat Rejection

The piston rests at bottom dead centre (BDC) for a moment and heat is rejected at constant volume ($V_4 = V_1$). In this process, pressure falls from p_4 to p_1, temperature decreases from T_4 to T_1 and entropy falls from S_4 to S_1.

The heat supplied, Q_s, per unit mass of charge is given by,

$$C_v(T_3 - T_4) \quad \text{...(i)}$$

The heat rejected, Q_r per unit mass of charge is given by,

$$C_v(T_4 - T_1) \quad \text{...(ii)}$$

The thermal efficiency is given by,

$$\eta_{th} = 1 - \frac{(T_4 - T_1)}{(T_3 - T_2)}$$

$$= 1 - \frac{T_1}{T_2}\left\{\frac{\left(\frac{T_4}{T_1} - 1\right)}{\left(\frac{T_3}{T_2} - 1\right)}\right\} \quad \text{...(iii)}$$

$$\text{Now, } \frac{T_1}{T_2} = \left(\frac{V_2}{V_1}\right)^{\gamma-1} = \left(\frac{V_3}{V_4}\right)^{\gamma-1} = \frac{T_4}{T_3}$$

Hence, substituting in Eq. 3, we get, assume r is the compression ratio $V_{1/}V_2$

$$\eta_{th} = 1 - \frac{T_1}{T_2}$$

$$=1-\left(\frac{V_2}{V_1}\right)^{\gamma-1}$$

$$\eta_{otto}=1-\frac{1}{(r)^{\gamma-1}}$$

If the value of compression 'r' increases, then the value of efficiency will increase correspondingly. Figure shown a plot of thermal efficiency versus compression ratio for an Otto cycle. It is seen that the increase in efficiency is significant at lower compression ratios.

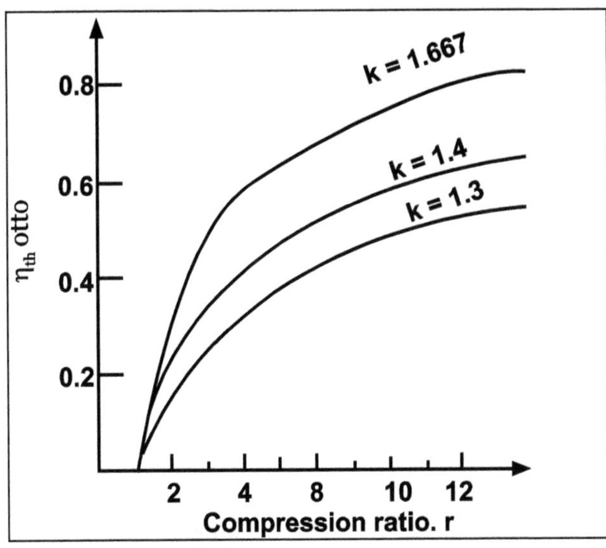

Variation of Efficiency with Compression Ratio.

Problems

1. Let us determine the thermal efficiency and compression ratio for an automobile working on Otto cycle. If the energy generated per cycle is thrice that of rejected during the exhaust. Consider working fluid as an ideal gas with $\gamma = 1.4$.

Solution:

Given:

$$\gamma=1.4$$

$$Q_1=3Q_2$$

To find:

The thermal efficiency

Formula to be used:

$$\eta_{otto}=Q_1-Q_2/Q_1$$

Where,

Q_1 = Heat supplied

Q_2 = Heat rejected

$$Q_1 = 3Q_2$$

$$\eta_{otto} = 3Q_2 - Q_2 / 3Q_2 = 2/3 = 66.67\%$$

We also have,

$$\eta_{otto} = 1 - 1/(r)^{\gamma-1}$$

$$0.6667 = 1/(r)^{1.4-1}$$

$$r = (3.00)^{1/0.4}$$

$$= 15.59$$

2. An engine works on Otto cycle. The initial pressure and temperature of the air is 1 bar and 40° C. 825 kJ of heat is supplied per kg of air at the end of the compression. Let us find the temperature and pressure at all salient points if the compression ratio is 6. Let us also determine the efficiency and mean effective pressure for the cycle assuming that air is used as working fluid and take all ideal conditions.

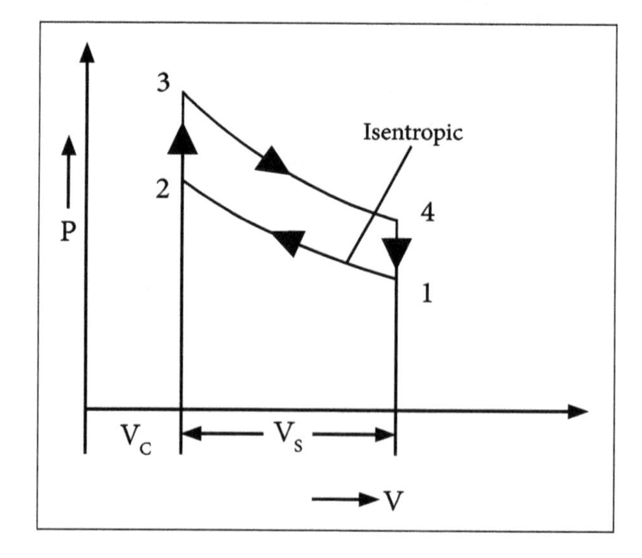

Solution:

Given:

Pressure P_1 = 1 bar = 100 kN/m²

T_1 = 40°C + 273

T_1 = 813 K

Q = 825 kJ

Compression ratio, y = 6

$$\left(\frac{V_1}{V_2} = \frac{V_4}{V_3}\right) = 6$$

To find:

- Efficiency (η).
- Work (w).

Formula to be used:

$$P_2 = \left(\frac{V_1}{V_2}\right)^{\gamma} \times P_1$$

$$\left(\frac{T_2}{T_1}\right) = \left(\frac{V_1}{V_2}\right)^{\gamma-1}$$

$$P_3 = \frac{T_3}{T_4} \times P_2$$

$$Q_S = mc_v\left(T_3 - T_2\right)$$

$$Q_R = mc_v\left(T_4 - T_1\right)$$

$$P_1V_1 = mRT_1$$

$$\rho_m = \frac{W}{V_1 - V_2}$$

$$W = Q_S - Q_r$$

Consider process 1 - 2 (adiabatic process):

$$\frac{P_2}{P_1} = \left(\frac{V_1}{V_2}\right)^{\gamma}$$

$$P_2 = \left(\frac{V_1}{V_2}\right)^{\gamma} \times P_1$$

$$= (6)^{1.4} \times 100$$

$$P_2 = 1228.6 \text{ kN/m}^2$$

Then,

$$\frac{T_2}{T_1} = \left(\frac{V_1}{V_2}\right)^{\gamma-1}$$

$$T_2 = (8)^{1.4-1} \times 313$$
$$T_2 = 719.08 \text{ k}$$

Consider process 2 – 3 (Constant volume process):

$$\frac{P_3}{P_2} = \frac{T_3}{T_2}$$

$$P_3 = \frac{T_3}{T_2} \times P_2$$

$$\frac{T_2}{T_1} = \left(\frac{V_1}{V_2}\right)^{\gamma-1}$$

$$Q_s = mC_v(T_3 - T_2)$$

$$\eta_{thermal} = \frac{W}{Q_S} = \frac{Q_S - Q_R}{Q_S}$$

$$Q_R = mC_v(T_4 - T_1)$$

$$P_m = P_1^{\gamma}\left(\frac{k-1}{\gamma-1}\right)\left(\frac{r^{\gamma-1}-1}{r-1}\right)$$

We know that, heat supplied,

$$Q_S = mc_v(T_3 - T_2)$$
$$825 = 1 \times 0.718 \times (T_3 - 719.08)$$
$$T_3 - 719.08 = 1149.025$$
$$T_3 = 186.08 \text{ k}$$

And the pressure,

$$P_3 = \frac{1868}{719.08} \times 1228.6$$
$$P_3 = 3191.92 \text{ kN/m}^2$$

Consider process 3 - 4 (adiabatic process):

$$\frac{P_4}{P_3} = \left(\frac{V_3}{V_4}\right)^{\gamma}$$

$$\frac{P_4}{3191.92} = (1/6)^{\gamma}$$

$$P_4 = 259.8 \text{ kN/m}^2$$

$$\frac{T_4}{T_3} = \left(\frac{V_3}{V_4}\right)^{\gamma-1}$$

$$\frac{T_4}{1868} = \left(\frac{1}{6}\right)^{1.4-1}$$

$$T_4 = 912.25 \text{ k}$$

We know that, heat rejected:

$$Q_R = mc_V(T_4 - T_1)$$

$$= 1 \times 0.718 \times (912.25 - 313)$$

$$Q_R = 430.26 \text{ kJ/kg}$$

The cycle efficiency,

$$\eta = 1 - \frac{1}{(r)^{\gamma-1}}$$

$$= 1 - \frac{1}{(y)^{1.4-1}}$$

$$= 0.5116$$

$$\eta = 51.16\%$$

The mean effective pressure:

$$P_1V_1 = mRT_1$$

$$100 \times V_1 = 1 \times 0.287 \times 313$$

$$V_1 = 0.89 \text{ M}^3/\text{kg}$$

We know that,

$$\frac{V_1}{V_2}=6 \Rightarrow \frac{0.89}{V_2}=6$$

$$V_2 = 0.149 \text{ m}^3/\text{kg}$$

$$\rho_m = \frac{W}{V_1 - V_2}$$

The work done,

$$W = Q_S - Q_r$$
$$= 825 - 430.26$$

$$W = 394.74 \text{ kJ/kg}$$

$$\therefore \text{ The } \rho_m = \frac{394.74}{0.149 - 0.89}$$

$$\rho_m = 532.71 \text{ kN/m}^2$$

3. In an Otto cycle, the air at the beginning of isentropic compression is at 1 bar and 15°c, The ratio of compression is 8. If the heat added during the constant volume process is 1000 kJ/kg. Let us determine:

- The maximum temperature in the cycle.
- The air standard efficiency.
- The work done per kg of air.
- The heat rejected.

Solution:

Given:

$T_1 = 15 + 273 = 288$ K

$P_1 = 1$ bar

$P_2 = 1$ bar

$V_1 / V_2 = 8$

$Q_3 = 1000$ kJ/kg

$\gamma = 1.4$

To find:

- The maximum temperature in the cycle.
- The air standard efficiency.
- The work done per kg of air.
- The heat rejected.

Formula to be used:

$$P_2 / P_1 = (V_1 / V_2)^\gamma$$
$$1 = -1 / r^{\gamma-1}$$

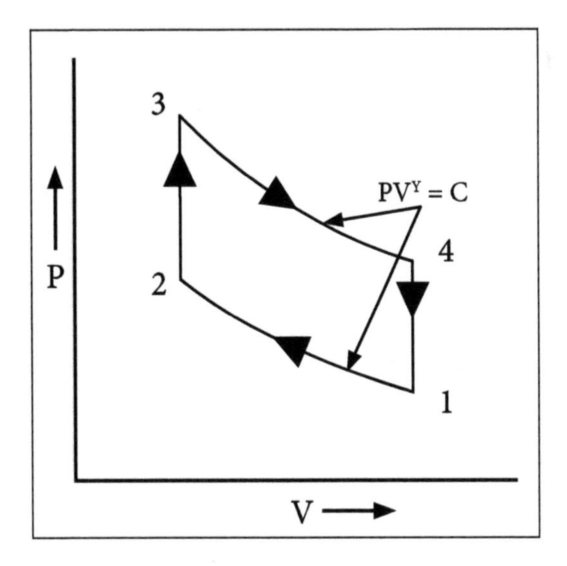

(a) Process (1-2): Isentropic Compression

$$P_2 / P_1 = (V_1 / V_2)^\gamma$$
$$P_2 = 1 \times (8)1.4 = 18.45 \text{ bar}$$

And also $T_2 / T_1 = (V_1 / V_2)^{\gamma-1}$

$$T_2 = 288\,(8)^{1.4-1} = 663 \text{ K}$$

$$\text{Heat supplied} = C_v(T_3 - T_2)$$

$$1000 = 0.718\,(T_3 - 663)$$
$$T_3 = 2056 \text{ K}$$

Maximum Temperature During the Cycle = 2056 K

(b) Air standard efficiency =1

$$= -1/r^{\gamma-1} = 1 - 1/(8)^{0.4}$$
$$= 0.565 \text{ (or } 56.5\%)$$

(c) Work done = Heat supplied X efficiency

$$= 1000 \times 0.565 = 565 \text{ kJ/kg}$$

(d) $T_3 / T_4 = 2056 / (8)^{1.4-1} = 894 \text{ K}$

$$T_4 = 2056 / (8)^{1.4-1} = 894 \text{ K}$$

$$\text{Heat rejected} = C_v(T_4 - T_1) = 0.718\ (894 - 288)$$

Heat rejected = 435 kJ/kg

4. For air standard Otto cycle the following data is available:

Compression ratio = 9,

Heat added/kg = 1200 kJ,

Lowest temperature in the cycle = 300 K,

Lowest pressure in the cycle = 1 bar,

Let us calculate the following:

- Pressure and temperature at each point in the cycle.
- Thermal efficiency.
- Mean effective pressure if air flow rate of, 0.25 kg/sec.

Assume C_p = 1 kJ/kgK and C_v = 0.714 kJ/kgK.

Solution:

Given:

Cycle → Otto cycle

$r = 9$

$Q_s = 1200$ kJ

$T_1 = 300$ k

$P_1 = 1$ bar = 100 kN/m²

m = 0.25 kg/sec

$C_P = 1$ kJ/kg-k

$C_V = 0.714$ kJ/kg-k

To Find:

- Pressure and temperature at each point of cycle.
- $\eta_{Thermal}$.
- P_m.

Formula to be used:

$$\frac{P_2}{P_1} = \left(\frac{V_1}{V_2}\right)^{\gamma}$$

$$\frac{T_2}{T_1} = \left(\frac{V_1}{V_2}\right)^{\gamma-1}$$

$$Q_s = mC_v\left(T_3 - T_2\right)$$

$$\eta_{thermal} = \frac{W}{Q_S} = \frac{Q_S - Q_R}{Q_S}$$

$$Q_R = mC_v\left(T_4 - T_1\right)$$

$$P_m = P_1^{\gamma}\left(\frac{k-1}{\gamma-1}\right)\left(\frac{r^{\gamma-1}-1}{r-1}\right)$$

Consider the Process 1–2. (Adiabatic Compression)

$$\frac{P_2}{P_1} = \left(\frac{V_1}{V_2}\right)^{\gamma}$$

$$P_2 = (9)^{1.4} \times 100$$
$$P_2 = 2167.40 KN/m^2$$

$$\frac{T_2}{T_1} = \left(\frac{V_1}{V_2}\right)^{\gamma-1}$$

$$T_2 = (9)^{0.4} \times 300$$
$$T_2 = 722.46k$$

Consider Process 2–3 (Constant Volume Process)

$$Q_s = mC_V(T_3 - T_2)$$

$$1200 = 0.25 \times 0.714\,(T_3 - 722.46)$$

$$T_3 = 7445.14\ k$$

Also, $\frac{P_3}{P_2} = \frac{T_3}{T_2}$,

$$P_3 = \frac{7445.14}{722.46} \times 2167.40$$

$$P_3 = 22335.62\ KN/m^2$$

Consider Process 3 – 4 (Adiabatic Process).

$$\frac{P_4}{P_3} = \left(\frac{V_3}{V_4}\right)^{\gamma}$$

$$P_4 = \left(\frac{V_3}{V_4}\right)^{\gamma} \times P_3$$

$$= \left(\frac{1}{9}\right)^{1.4} x22335.62$$

$$P_4 = 1030.52KN/m^2$$

$$\frac{T_4}{T_3} = \left(\frac{V_3}{V_4}\right)^{\gamma-1}$$

$$T_4 = (1/9)^{0.4} \times 7445.14$$

$$T_4 = 3091.54\ k$$

$$Q_2 = 498.28$$

$$\eta_{thermal} = \frac{W}{Q_S} = \frac{Q_S - Q_R}{Q_S}$$

$$Q_R = mC_v(T_4 - T_1)$$
$$= 0.25x0.714(3091.54 - 300)$$

$$\eta_{thermal} = \frac{1200 - 498.28}{1200}$$

$$= 58.47\%$$

$$P_m = P_1^{\gamma}\left(\frac{k-1}{\gamma-1}\right)\left(\frac{r^{\gamma-1}-1}{r-1}\right)$$

$$k = \frac{P_3}{P_2}\left(\frac{22335.62}{2167.40}\right) = 10.30$$

$$P_m = 100x9\left(\frac{10.30-1}{1.4-1}\right)\left(\frac{9^{0.4}-1}{9-1}\right)$$

$$P_m = 3683.38\ N/m^2$$

Result:

$$P_2 = 2167.40\ kN/m^2$$
$$P_3 = 22335.62\ kN/m^2$$
$$P_4 = 1030.52\ kN/m^2$$
$$T_2 = 722.46K$$
$$T_3 = 7445.14K$$
$$T_4 = 3091.54K$$

$$\eta_{thermal} = 58.47\%$$

$$P_m = 3683.38\ kN/m^2$$

2.1.1 Diesel Cycle

Diesel Cycle: This is the cycle, which was introduced by Rudolph diesel. This cycle is used in Diesel engines. It consists of the following four processes:

- Two adiabatic (or) isentropic process.
- One constant volume process.
- One constant pressure process.

The p - V and T - s diagram of diesel cycle are shown in the figure below:

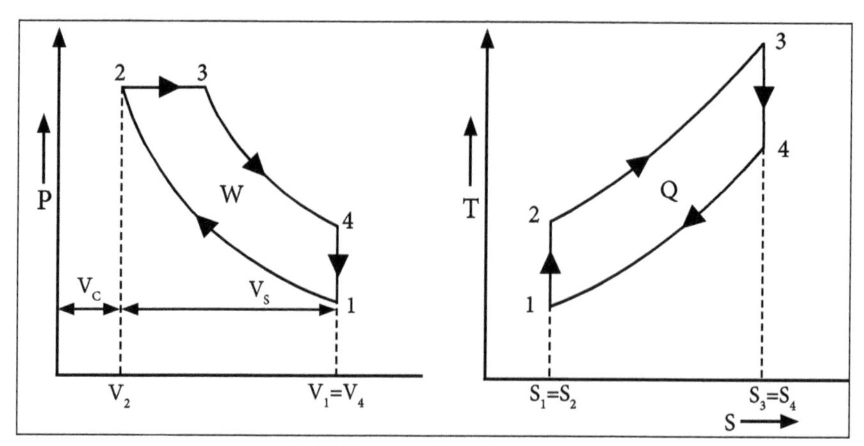

P- V diagram T - s diagram.

Process 1-2: During this process, air is compressed isentropically from P_1 to P_2 entropy remains constant ($S_1 = S_2$).

Process 2-3: During this is process air is heated from T_2 to T_3 but Pressure remain constant ($P_2 = P_3$),

Heat supplied during the process is given by,

$$Q_S = m \times C_p (T_3 - T_2)$$

Process 3-4: During this process, air expands isentropically from P_3 to P_4 Temperature decreases from to T_3 to T_4.

Process 4-1: During this process, heat is rejected from air but volume remains constant. Thus the temperature decreases from T_4 to T_1. Heat rejected during the process is given by,

$$Q_r = m \times C_v (T_4 - T_1)$$

Efficiency of Diesel Cycle

$$\eta_{Diesel} = \frac{Q_S - Q_R}{Q_S}$$

$$= \frac{mC_p (T_3 - T_2) - mC_V (T_4 - T_1)}{mC_p (T_3 - T_2)}$$

$$= 1 - \frac{mC_V (T_4 - T_1)}{mC_p (T_3 - T_2)} \left[\because \frac{C_P}{C_V} = \gamma\right]$$

$$\eta_{Diesel} = 1 - \frac{(T_4 - T_1)}{\gamma x (T_3 - T_2)}$$

The efficiency is in terms of temperature only. Hence, equation is simplified in terms of volume ratio.

$$\text{Compression ratio} = \frac{\text{Total cylinder volume}}{\text{Clearance volume}} = \frac{V_1}{V_2} = r$$

The cut off ratio is the ratio between volume at the point of cut-off and clearance volume,

$$\text{Cut-off ratio } \rho = \frac{\text{Cut-off Volume}}{\text{Clearance Volume}} = \frac{V_3}{V_2}$$

$$\text{Compression ratio} = \frac{\text{Total cylinder volume}}{\text{Clearance volume}} = \frac{V_1}{V_2} = r$$

Consider Process 1-2

From adiabatic relation,

$$\frac{T_2}{T_1} = \left[\frac{V_1}{V_2}\right]^{\gamma-1}$$

$$\frac{T_2}{T_1} = (r)^{\gamma-1}$$

$$T_2 = T_1 \times r^{\gamma-1}$$

Consider Process 2-3

Process 2-3 is a Constant Pressure Process $\frac{V}{T} = C$.

$$\frac{V_2}{T_2} = \frac{V_3}{T_3}$$

$$\frac{T_3}{T_2} = \frac{V_3}{V_2} = \rho$$

$$T_3 = T_2 \text{ x } \rho = T_1 \text{ x } (r)^{\gamma-1} \rho \; (\therefore \; T_2 = T_1 (r)^{\gamma-1})$$

$$T_3 = T_1 \text{ x } (r)^{\gamma-1} \rho$$

Consider Process 3-4

Using adiabatic equation,

$$\frac{T_3}{T_4} = \left[\frac{V_4}{V_3}\right]^{\gamma-1} = \left[\frac{r}{\rho}\right]^{\gamma-1}$$

$$T_4 = \frac{T_3}{\left(\frac{r}{\rho}\right)^{\gamma-1}} = \frac{T_1 (r)^{\gamma-1} \rho}{\left(\frac{r}{\rho}\right)^{\gamma-1}}$$

$$T_4 = \frac{T_1 (r)^{\gamma-1} \rho \rho^{\gamma-1}}{(r)^{\gamma-1}}$$

$$T_4 = T_1 \rho^{\gamma}$$

Substitute T_2, T_3, T_4 values in η_{diesel} equation,

$$\eta_{diesel} = 1 - \frac{1}{\gamma}\left[\frac{T_1\rho^{\gamma} - T_1}{T_1(r)^{\gamma-1}\rho - T_1(r)^{\gamma-1}}\right]$$

$$= 1 - \frac{1}{\gamma}\left[\frac{T_1\left(\rho^{\gamma} - 1\right)}{T_1 r^{\gamma-1}(\rho - 1)}\right]$$

$$\eta_{diesel} = 1 - \frac{1}{\gamma(r)^{\gamma-1}}\left[\frac{\rho^{\gamma-1}}{\rho - 1}\right].$$

From the above equation,

- If the compression ratio increases and efficiency of the diesel cycle increases and vice versa.
- The efficiency of diesel cycle decreases with increase in cut-off ratio and vice versa.

Problems

1. For air standard diesel cycle the following data is available; Compression ratio = 16, Heat added/ kg = 2500 kJ/kg, Lowest temperature in the cycle = 300 K, Lowest pressure in the cycle = 1 bar, Let us calculate the:

- Pressure and temperature at each point in the cycle,
- Thermal efficiency,
- Mean effective pressure if air flow rate of 0.25 kg/sec.

Assume Cp = 1 kJ/kg and Cv = 0.714 kJ/kgK.

Given:

Cycle → Direct Cycle

Solution:

Compression ratio $r = \frac{V_1}{V_2} = 16$,

$m = 0.25$ kg/sec

$Q_s = 2500$ kJ/kg

$T_1 = 300$ k

$P_1 = 1$ bar $= 100$ kN/m²

$C_p = 1$ kJ/kg-k

$C_v = 0.714$ kJ/kg-k

To Find:

- Pressure and temperature at each point of cycle.
- $\eta_{Thermal.}$
- $p_{m.}$

Formula to be used:

$$\frac{T_2}{T_1} = \left(\frac{V_1}{V_2}\right)^{\gamma-1}$$

$$\frac{T_2}{T_1} = \left(\frac{P_2}{P_1}\right)^{\gamma-1/\gamma}$$

$$Q_s = mC_p\left(T_3 - T_2\right) = 2500\ KJ/Kg$$

$$\eta_{thermal} = \frac{W}{Q_S} = \frac{Q_S - Q_R}{Q_S}$$

$$P_m = \frac{P_1 r^{\gamma}\left[\gamma(\rho-1) - r^{1-\gamma}\left(\rho^{\gamma}-1\right)\right]}{(\gamma-1)(r-1)}$$

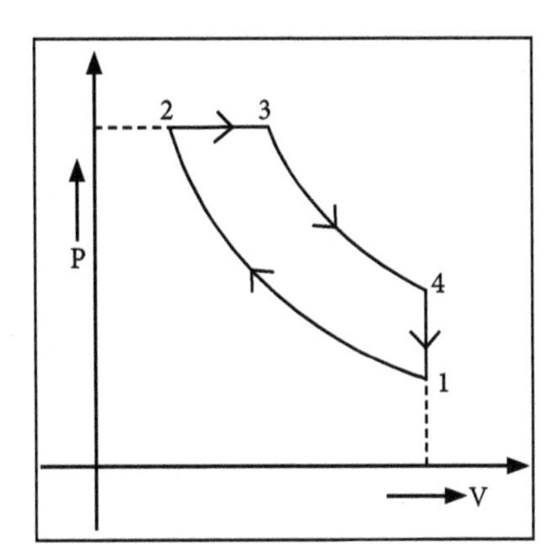

$$\frac{T_2}{T_1} = \left(\frac{V_1}{V_2}\right)^{\gamma-1}$$

$$T_2 = 300(16)^{1.4-1}$$
$$T_2 = 909.24k$$

$$\frac{T_2}{T_1} = \left(\frac{P_2}{P_1}\right)^{\gamma-1/\gamma}$$

$$P_2 = \left(\frac{T_2}{T_1}\right)^{\gamma/\gamma-1} x\ P_1 = \left(\frac{909.42}{300}\right)^{1.4/0.4}$$

$$P_2 = 4850.29KN/m^2$$
$$P_3 = P_2 = 4850.29KN/m^2$$

Consider the Process 2-3 (Constant Pressure Process)

$$Q_s = m \times C_P(T_3 - T_2) = 2500\ kJ/kg$$
$$P\ 2500 = 0.25 \times 1(T_3 - 909.42)$$
$$P\ T_3 = 10909.42\ k$$

$$\frac{1/2}{T_2} = \frac{V_3}{T_3} \Rightarrow \frac{V_3}{V_2} = \frac{T_3}{T_2} = \frac{10909.42}{909.42} = 11.996$$

$$\text{Cut - off ratio, } \rho = \frac{V_3}{V_2} = 11.996$$

Consider the Process 3 - 4 (Adiabatic Expansion)

$$\frac{T_4}{T_3} = \left(\frac{V_3}{V_4}\right)^{\gamma-1}$$

$$T_4 = \left(\frac{V_3}{V_2} x \frac{V_2}{V_1}\right)^{\gamma-1} x\ T_3 (\because V_4 = V_1)$$

$$= \left(\frac{11.996}{16}\right)^{0.4} x\ 10909.42$$

$$T_4 = 9722.28k$$

$$\frac{T_4}{T_3} = \left(\frac{P_4}{P_3}\right)^{\gamma-1/\gamma}$$

$$P_4 = \left(\frac{T_4}{T_3}\right)^{\gamma/\gamma-1} x\ P_3$$

$$= \left(\frac{9722.28}{10909.42}\right)^{1.4/0.4} x\ 4850.29$$

$$P_4 = 3240.78 KN/m^2$$

$$\eta_{thermal} = \frac{W}{Q_S} = \frac{Q_S - Q_R}{Q_S}$$

$$Q_R = m\ x\ C_V (T_4 - T_1) = 0.25\ x\ 0.714(9722.28 - 300)$$

$$Q_R = 1681.87 KJ$$

$$\eta_{thermal} = \frac{2500 - 1681.87}{2500}$$

$$= 32.72\%$$

Mean Effective Pressure

$$P_m = \frac{P_1 r^\gamma \left[\gamma(\rho - 1) - r^{1-\gamma}(\rho^\gamma - 1)\right]}{(\gamma - 1)(r - 1)}$$

$$= \frac{100\ x 16^{14} \left[1.4(11.996 - 1) - 16^{1-1.4}(11.996^{1.4} - 1)\right]}{(1.4 - 1)(16 - 1)}$$

$$P_m = 4069.11 KN/m^2$$

Result:

$P_2 = 4850.29$ kN/m²

$P_3 = 4850.29$ kN/m²

$P_4 = 3240.78$ kN/m²

$T_2 = 909.42$ K

$T_3 = 10909.42$ K

$T_4 = 9722.28$ K

$\eta_{Thermal} = 32.72\%$

$P_m = 4069.11$ KN/m²

2. For a diesel cycle, the compression ratio is 15. Compression begins at 0.1 Mpa, 40°C. The heat added is 1.675 MJ/kg. Let us determine (a) the maximum temperature in the cycle, (b) work done per kg of air (c) the cycle efficiency (d) the temperature at the end of the isentropic expansion (e) the cut-off ratio and (f) the MEP of the cycle.

Solution:

Given:

T_1=40°C=313 K

P_1=0.1Mpa

Q_{in}=1675 MJ/kg

$r = v_1/v_2$=15

To find:

- The maximum temperature in the cycle.
- Work done per kg of air.
- The cycle efficiency.
- The temperature at the end of the isentropic expansion.
- The cut-off ratio.
- The mep of the cycle.

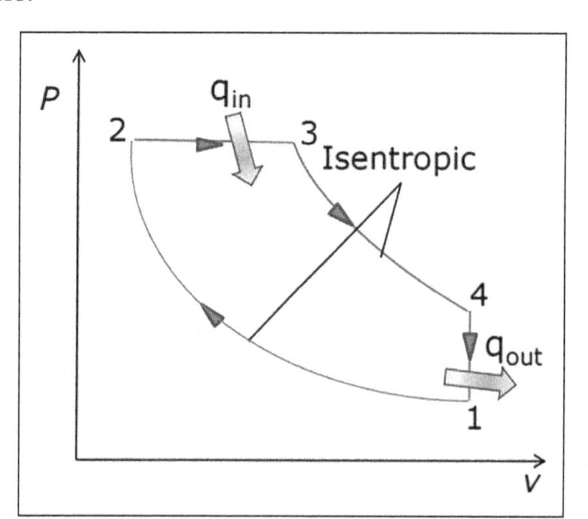

Formula to be used:

$$Q_{in} = C_p\left(T_3 - T_2\right)$$

$$V_1 = \frac{RT_1}{P_1}$$

$$V_2 = V_1/15$$

$$T_4 = T_3\left(\frac{V_3}{V_4}\right)$$

$$v_1 = \frac{RT_1}{P_1} = \frac{0.287 \text{ x } 313}{100} = 0.898\,m^3/kg$$

$$V_2 = v_1/15 = 0.898/15 = 0.06m^3/kg$$

It is given that $Q_{in} = 1675\ MJ/kg$,

$$Q_{in} = c_p\left(T_3 - T_2\right)$$

$$\frac{T_2}{T_1} = \left(\frac{v_1}{v_2}\right)^{\gamma-1} = 15^{0.4} = 2.954$$

$$T_2 = 313 \text{ x } 2.954 = 924.66K$$
$$Q_{in} = 1675 = 1.005\left(T_3 - 924.66\right)$$

$$\therefore T_3 = 2591.33\ K = T_{max}$$

Hence, the maximum temperature is 2591.33 K.

$$\frac{P_2}{P_1} = \left(\frac{v_1}{v_2}\right)^{\gamma} = 15^{1.4} = 44.31$$

$$\therefore p_2 = 4431\ kPa$$

$$\frac{P_2 v_2}{T_1} = \frac{P_3 v_3}{T_3} \rightarrow v_3 = \frac{T_3}{T_2} v_2 = \frac{2591.33}{924.66} \text{ x } 0.06 = 0.168\ m^3/kg$$

$$r_c = \frac{v_3}{v_2} = \frac{0.168}{0.06} = 2.8$$

The cut-off ratio is 2.8.

$$T_4 = T_3\left(\frac{v_3}{v_4}\right)^{\gamma-1} = 2591.33 \text{ x } \left(\frac{0.168}{0.898}\right)^{0.4}$$
$$= 1325.37\ K$$

$$Q_{out} = C_v\left(T_4 - T_1\right) = 0.718\left(1325.4 - 313\right) = 726.88\,kJ/kg$$

Net Work Done, $W_{net} = Q_{in} - Q_{out} = 1675 - 716.88$

$= 948.12 \text{ kJ/kg}$

2.1.2 Dual

Cycle

The cycle is the equivalent air cycle for reciprocating high speed compression ignition engines. The P-V and T-s diagrams are shown in figure given below. In the cycle, compression and expansion processes are isentropic; heat addition is partly at constant volume and partly at constant pressure while heat rejection is at constant volume as in the case of the Otto and Diesel cycles.

The dual combustion, cycle consists of the following operations:

- Adiabatic compression (1-2).
- Addition of heat at constant volume (2-3).
- Addition of heat at constant pressure (3-4).
- Adiabatic expansion (4-5)..
- Rejection of heat at constant volume (5-1).

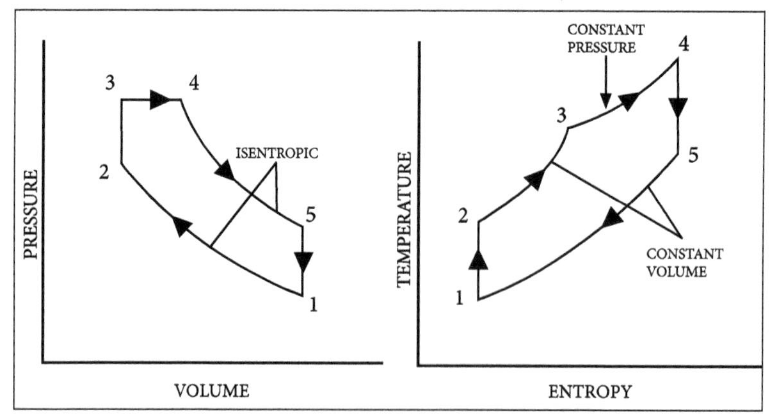

Dual combustion cycle.

Consider 1 kg of air,

Total heat supplied = Heat supplied during the operation 2–3 + heat supplied during the operation 3–4

$= C_v(T_3 - T_2) + c_p(T_4 - T_3)$

Heat Rejected During Operation 5–1 = $c_v(T_5 - T_1)$

Work done = Heat supplied − heat rejected.

$$= C_b(T_3 - T_2) + c_p(T_4 - T_3) - c_\upsilon(T_5 - T_1)$$

$$\mu_{dual} = \frac{\text{Work done}}{\text{Heat Supplied}} = \frac{c_v(T_3 - T_2) + c_p(T_4 - T_3) - cv(T_5 - T_1)}{c_v(T_3 - T_2) + c_p(T_4 - T_3)}$$

$$= 1 - \frac{c_v(T_5 - T_1)}{c_v(T_3 - T_2) + c_p(T_4 - T_3)}$$

$$= 1 - \frac{c_v(T_5 - T_1)}{(T_3 - T_2) + \gamma(T_4 - T_3)} \qquad \left(\because \gamma = \frac{c_p(i)}{c_v}\right) \text{(i)}$$

Compression ratio, $r = \frac{V_1}{V_2}$

During Adiabatic Compression Process 1–2,

$$\frac{T_2}{T_1} = \left(\frac{V_1}{V_2}\right)^{\gamma-1} = (r)^{\gamma-1}$$

During Constant Volume Heating Process

$$\frac{P_3}{T_3} = \frac{P_2}{T_2}$$

$$\frac{T_3}{T_2} = \frac{P_3}{P_2} = \beta$$

Where, β is known as pressure or explosion ratio.

or T —

During adiabatic expansion process,

$$\frac{T_4}{T_5} = \left(\frac{V_5}{V_4}\right)^{\gamma-1}$$

$$= \left(\frac{r}{\rho}\right)^{\gamma-1}$$

$$\left(\because \frac{V_5}{V_4} = \frac{V_1}{V_4} = \frac{V_1}{V_2} \times \frac{V_2}{V_4} = \frac{V_1}{V_2} \times \frac{V_3}{V_4} = \rho \text{ being the cut-off ratio}\right)$$

During constant pressure heating process,

$$\frac{V_3}{T_3} = \frac{V_4}{T_4}$$

$$T_4 = T_3 \frac{V_4}{V_3} = \rho T_3 \qquad \text{...(ii)}$$

Putting the value of T_4 in the equation (i), we get,

$$\frac{\rho T_3}{T_5} = \left(\frac{r}{\rho}\right)^{\gamma-1} \text{ or } T_5 = \rho . T_3 . \left(\frac{\rho}{r}\right)^{\gamma-1}$$

Putting the value of $.T_2$ in equation (ii) we get,

$$\frac{\frac{T_3}{\beta}}{T_1} = (r)^{\gamma-1}$$

$$T_1 = \frac{T_3}{\beta} \cdot \frac{1}{(r)^{\gamma-1}}$$

Now inserting the values of T_1, T_2, T_4 and T_5 in equation,

$$\eta_{dual} = 1 - \frac{\left[\rho \cdot T_3 \left(\frac{\rho}{r}\right)^{\gamma-1} - \frac{T_3}{\beta} \cdot \frac{1}{(r)^{\gamma-1}}\right]}{\left[\left(T_3 - \frac{T_3}{\beta}\right) + \gamma\left(\rho T_3 - T_3\right)\right]} = 1 - \frac{\frac{1}{(r)^{\gamma-1}}\left(\rho^{\gamma} - \frac{1}{\beta}\right)}{\left(1 - \frac{1}{\beta}\right) + \gamma(\rho - 1)}$$

$$\eta_{dual} = 1 - \frac{1}{(r)^{\gamma-1}} \cdot \frac{\left(\beta \cdot \rho^{\gamma} - 1\right)}{\left[(\beta - 1) + \beta\gamma(\rho - 1)\right]}$$

Problems

1. The compression ratio of an air standard dual cycle is 12 and the maximum pressure in the cycle is limited to 70 bar. The pressure and temperature of the cycle at the beginning of compression process are 1 bar and 300K. Let us calculate the thermal efficiency and mean effective pressure assuming Cylinder bore = 250 mm, stroke length = 300 mm, C_p = 1.005 kJ/kgK, C_v = 0.718 kJ/kgK. Heat is added during constant process up to of the stroke.

Solution:

Given data:

$r = 12$

$P_3 = P_4 = 70 \times 10^3\ N/m^2$

$P_1 = 1 \times 10^5\ N/m^2$

$T_1 = 300\ k$

$\frac{\rho-1}{\gamma-1} = 0.03$

$D = 0.25\ m$

Stroke $L = 0.3\ m$

Formula to be used:

$$\eta_{Dual} = 1 - \frac{1}{(r)^{\gamma-1}}\left[\frac{\gamma_p e^{\gamma} - 1}{(r_p - 1) + \gamma r_p(\rho - 1)}\right]$$

$$Q_s = Q_{sV} + Q_{sp}$$
$$Q_s = mC_v(T_3 - T_2) + mC_p(T_4 - T_3)$$

Heat rejected $Q_R = m\, C_v(T_5 - T_1)$,

$$MEP = \frac{Workdone}{Stroke\ Volume} = \frac{Q_S - Q_R}{V_S}$$

$$gn_1\ \frac{\rho-1}{\gamma-1} = 0.03$$

$$\rho = 0.03\ (12-1) + 1 = 1.33$$
$$Cut-off\ ratio\ \rho = 1.33$$

Cycle Efficiency

$$\eta_{Dual} = 1 - \frac{1}{(r)^{\gamma-1}}\left[\frac{\gamma_p e^{\gamma} - 1}{(r_p - 1) + \gamma r_p(\rho - 1)}\right]$$

Heat Supplied

$Q_s = Q_{sV} + Q_{sp}$ = {Heat supplied at constant volume} + {Heat supplied at Constant pressure}

$$Q_s = mC_v(T_3 - T_2) + mC_p(T_4 - T_3)$$

$$m = \frac{\rho_1 V_1}{RT_1} = \frac{1 \text{ x } 10^5 \text{ x } 0.016}{287 \text{ x } 300}$$

$$M = 0.0186 \text{ kg}$$

$$\therefore Q_s = 0.0186 \left(718 \left(1750.03 - 810.576\right) + 1005\left(2327.54 - 1750.03\right)\right]$$

$$Q_s = 23.34 \times 10^3 \text{W}$$

$$\text{Heat rejected } Q_R = mC_v\left(T_5 - T_1\right)$$

$$= 0.186 \times T_{18}\left(\gamma_p \rho^\gamma T_1 - T_1\right)$$

$$= 0.0186 \times 718 \times 300\left(2.159 \times 1.33^{1.4} - 1\right)$$

$$Q_R = 8888.03 \text{ W}$$

Mean Effective Pressure,

$$\text{MEP} = \frac{\text{Workdone}}{\text{Stroke Volume}}$$

$$= \frac{Q_S - Q_R}{Q_S}$$

$$= \frac{23340 - 8888.03}{0.0147}$$

$$\text{MEP}, P_m = 9.8312 \text{ bar}$$

2. An oil engine working on dual cycle, the heat supplied at constant pressure is twice that of heat supplied at constant volume. The compression and expansion ratios are 8 and 5.3. The pressure and temperature at the beginning of cycle are 0.93 bar and 27° C. Let us calculate the efficiency of the cycle and mean effective pressure taking C_p = 1.005 kJ/kgK and C_v = 0.718 kJ/kgK.

Solution:

Given:

Dual Cycle,

Q_s at constant pressure = $2Q_S$ at constant volume.

Compression ratio $r_c = 8$

Expansion ratio $r_e = 5.3$

P_1 = 0.93 bar = 93 kN/m²

T_1=27°C +273 = 300 k.

C_p = 1.005 kJ/kgk,

C_v = 0.718 kJ/kgk

To Find:

- Efficiency of the cycle.
- Mean effective pressure.

Formula to be used:

$$P_2 = P_1 \text{ x} \left[\frac{V_1}{V_2}\right]^{\gamma}$$

$$T_2 = T_1 \text{ x} \left[\frac{V_1}{V_2}\right]^{\gamma-1}$$

$$\frac{T_4}{T_3} = \frac{V_4}{V_3} = \rho = \frac{(\text{Compression ratio}(r_c))}{\text{Expansion ratio}(r_e)}$$

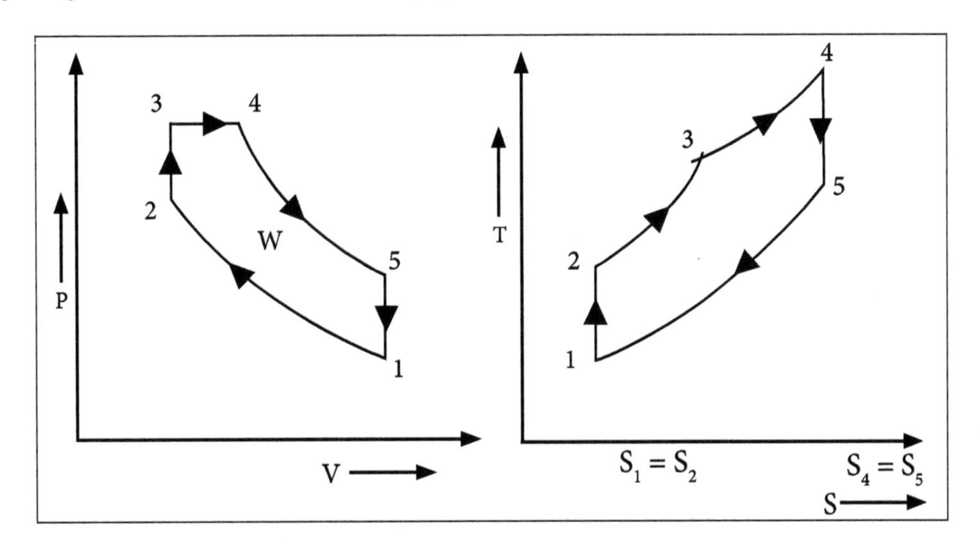

Isentropic Compression Process

$$P_1V_1^{\gamma} = P_2V_2^{\gamma}$$

$$P_2 = P_1 \text{ x} \left[\frac{V_1}{V_2}\right]^{\gamma}$$

$$= 93\text{x}(8)^{1.4} = 1709.26 \text{ KN} / \text{m}^2$$

$$P_2 = 17.09 \text{ bar}$$

$$\frac{T_2}{T_1} = \left[\frac{V_1}{V_2}\right]^{\gamma-1}$$

$$T_2 = T_1\left[\frac{V_1}{V_2}\right]^{\gamma-1}$$

$$T_2 = T_1 \times [8]^{1.4-1}$$
$$= 300 \text{ x } (8)^{1.4-1}$$
$$T_2 = 689.21\text{k}$$
$$C_p(T_4 - T_3) = 2 \text{ x } C_\upsilon(T_3 - T_2) \quad \text{...(i)}$$

Given Constant Pressure Process 3-4,

$$\frac{\rho_3 V_3}{T_3} = \frac{\rho_4 V_4}{T_4}$$

$$\frac{T_4}{T_3} = \frac{V_4}{V_3} = \rho = \frac{(\text{Compression ratio}(r_c))}{\text{Expansion ratio}(r_e)}$$

$$\rho = \text{Cut - off ratio}$$

$$\rho = \frac{r_c}{r_e} = \frac{8}{5.3}$$

$$\rho = 1.50943$$

$$\frac{T_4}{T_3} = \frac{V_4}{V_3} = \rho$$

$$\frac{T_4}{T_3} = \rho$$

$$\frac{V_5}{V_4} = r_e = \frac{V_5}{V_3} \text{ x } \frac{V_3}{V_4}$$

$$(\because V_5 = V_1)$$

$$\frac{V_4}{V_3} = \rho$$

$$= \frac{V_1}{V_3} x \frac{1}{\rho}$$

$$(\because V_3 = V_2)$$

$$= \frac{V_1}{V_2} x \frac{1}{\rho}$$

$$\frac{T_4}{T_3} = 1.50943$$

$$r_e = r_c \, x \frac{1}{\rho}$$

$$T_4 = 1.50943 \, T_3$$

$$\delta = \frac{r_c}{r_e} = \frac{\frac{r_c}{V_5}}{V_4}$$

Substituting Values of T_2 and T_4 in Equation,

$$c_p\left(T_4 - T_3\right) = 2 \times C_v\left(T_3 - T_2\right)$$
$$1.005\left[1.50943\,T_3 - T_3\right] = 2 \times Cv\left(T_3 - 689.21\right)$$
$$1.51697\,T_3 - 1.005\,T_3 = 2 \times 0.718\left(T_3 - 689.21\right)$$
$$1.51697\,T_3 - 1.005\,T_3 = 1.436\left(T_3 - 689.21\right)$$
$$0.51197\,T_3 = 1.436\,T_3 - 989.7055$$

$$989.7055 = 1.436\,T_3 - 0.51197\,T_3$$
$$989.7055 = \left(1.463 - 0.51197\right) T_3$$
$$989.7055 = 0.92403\,T_3$$
$$T_3 = 1071.07 \text{ k}$$

2-3 Constant Volume Process,

$$\frac{P_3}{T_3}=\frac{P_2}{T_2}$$

$$P_3=\frac{P_2}{T_2}\times T_3$$

$$P_3=P_2\times\left(\frac{T_3}{T_2}\right)$$

$$=1709.26\times\frac{1071.07}{689.21}$$

$P_3 = 2656.28\ KN/m^2$ or 26.56 bar

$P_4 = P_3 = 2656.28\ KN/m^2$

$T_4 = 1.50943T_3$

$T_4 = 1.5094 \times 1071.07$

$T_4 = 1616.70\ k$

Process 4-5

$$P_4V_4^{\gamma} = P_5V_5^{\gamma}$$

$$P_5=P_4\times\left(\frac{V_4}{V_5}\right)^{\gamma}$$

$$P_5=P_4\times\frac{1}{(r_e)^{\gamma}}=\frac{P_4}{(r_e)^{\gamma}}=\frac{2656.28}{(5.3)^{1.4}}$$

$$P_5 = 257.209 kN/m^2$$

$$\frac{T_5}{T_4}=\left[\frac{V_4}{V_5}\right]^{\gamma-1}=\frac{1}{(\gamma_c)^{\gamma-1}}=\frac{1}{(5.3)^{1.4-1}}=0.5132$$

$T_5 = T_4 \times 0.5132$

$=1616.70 \times 0.5132$

$T_5 = 829.69\ k$

$W = Q_s - Q_R$

$Qs = Q_s$ constant volume + Q_s constant pressure

= Q_s constant volume + 2 Q_s constant volume

= $3Q_s$ constant volume

$= 3C_v(T_3 - T_2)$

$= 3 \times 0.718 \times [1071.07 - 689.21]$

$Q_s = 822.526$ kJ

$Q_R = mC_v(T_5 - T_1)$

$= 1 \times 0.718 \times [829.69 - 300]$

$Q_R = 380.31$ kJ

$$\eta_{dual} = \frac{Q_S - Q_R}{Q_S} = \frac{822.526 - 280.31}{822.526}$$

$= 0.5376$

$\eta_{dual} = 53.76\%$

$$\eta = \frac{W}{Q_S}$$

$W = \eta \times Q_s = 0.5376 \times 822.526$

$W = 442.189$ KJ

$$P_m = \frac{W}{V_1 - V_2}$$

$P_1V_1 = mRT_1$

$P_1V_1 = P_2V_2$

$93 \times V_1 = 0.287 \times 300$

$93V_1 = 86.1$

$$V_1 = \frac{86.1}{93}$$

$V_1 = 0.9258\ m^3 / Kg$

$$\frac{V_1}{V_2} = 8$$

$$V_2 = 0.115725 \text{ m}^3/\text{kg}$$

$$P_m = \frac{W}{V_1 - V_2} = \frac{442.189}{0.9258 - 0.115725} = 545.86 \text{ kN}/\text{m}^2$$

$$P_m = 545.86 \text{KN}/\text{m}^2 \text{or } P_m = 5.45 \text{ bar}$$

3. An ideal dual cycle has a compression ratio of 14 and cutoff ratio of 1.2. Let us determine the air standard efficiency.

Solution:

Given:

$$c_p = 1.005 \text{ kJ}/\text{kg·K}$$
$$c_v = 0.718 \text{ kJ}/\text{kg·K}$$
$$R = 0.287 \text{ kJ}/\text{kg·K}$$
$$k = 1.4$$

To find:

Air Standard efficiency

Formula to be used:

- $v_1 = \frac{RT_1}{P_1}$
- $v_2 = \frac{v_1}{r}$
- $P_x = P_3 = r_p P_2$
- $q_{in} = c_v(T_x - T_2) + c_p(T_3 - T_x)$
- $\eta_{th} = 1 - \frac{q_{out}}{q_{in}}$

The specific volume of the air at the start of the compression is given by,

$$v_1 = \frac{RT_1}{P_1} = \frac{(0.287 \text{kPa.m}^3/\text{kg.K})(253\text{K})}{80 \text{kPa}} = 0.9076 \text{m}^3/\text{kg}$$

The specific volume at the end of the compression is given by,

$$v_2 = \frac{v_1}{r} = \frac{0.9076 \text{m}^3/\text{kg}}{14} = 0.06483 \text{m}^3/\text{kg}$$

The pressure at the end of the compression is given by,

$$P_2 = P_1\left(\frac{v_1}{v_2}\right)^k = P_1 r^k = (80\text{kPa})(14)^{1.4} = 3219\text{kPa}$$

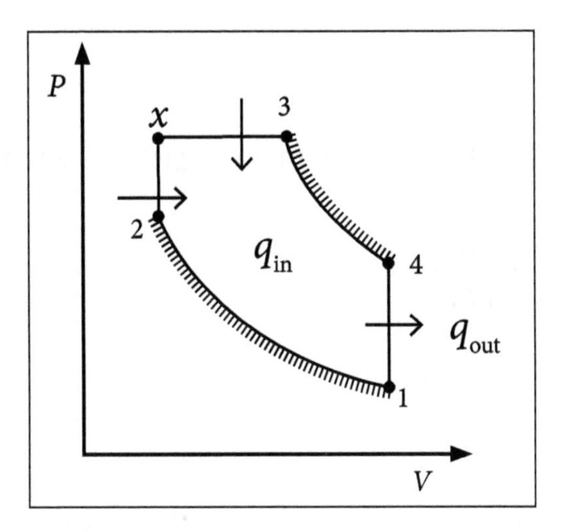

The maximum pressure is given by,

$$P_x = P_3 = r_p P_2 = (1.5)(3219\text{KPa})$$
$$= 4829\text{kpa}$$

The temperature at the end of the compression is given by,

$$T_2 = T_1\left(\frac{v_1}{v_2}\right)^{k-1} = T_1 r^{k-1} = (253\text{K})(14)^{1.4-1} = 727.1\text{K}$$

And

$$T_x = T_2\left(\frac{P_3}{P_2}\right) = (727.1\text{K})\left(\frac{4829\text{ kPa}}{3219\text{ kPa}}\right) = 1091\text{K}$$

Cutoff ratio is given,

$$v_3 = r_c v_x = r_c v_2 = (1.2)(0.06483\text{m}^3/\text{kg}) = 0.07780\text{m}^3/\text{kg}$$

The remaining state temperatures are then,

$$T_3 = T_x\left(\frac{v_3}{v_x}\right) = (1091\text{K})\left(\frac{0.07780}{0.9076}\right) = 1309\text{ K}$$

$$T_4 = T_3\left(\frac{v_3}{v_4}\right)^{k-1} = (1309\text{K})\left(\frac{0.07780}{0.9076}\right)^{1.4-1} = 490.0\text{ K}$$

Applying the first law work expression to the heat addition processes given by,

$$q_{in} = c_v\left(T_x - T_2\right) + c_p\left(T_3 - T_x\right)$$
$$= \left(0.718KJ/Kg.K\right)\left(1091 - 727.1\right)K + \left(1.005KJ/Kg.K\right)\left(1309 - 1091\right)K$$
$$= 480.4\ kJ/kg$$

The heat rejected is given by,

$$q_{out} = c_v\left(T_4 - T_1\right) = \left(0.718KJ/Kg.K\right)\left(490.0 - 253\right)K = 170.2KJ/Kg$$

$$\eta_{th} = 1 - \frac{q_{out}}{q_{in}} = 1 - \frac{170.2\,kJ/kg}{480.4\,kJ/kg}$$

$$\eta_{th} = 0.646 = 64\%$$

2.2 Description and Operation of Four and Two Stroke Cycle Engine

Petrol Engines

In this gasoline is mixed with air, broken up into a mist and partially vaporized in a carburetor. The mixture is then sucked into the cylinder. There it is compressed by the upward movement of the piston and is ignited by an electric spark.

When the mixture is burned, the resulting heat causes the gases to expand. The expanding gases exert a pressure on the piston (power stroke). The exhaust gases escape in the next upward movement of the piston. The compression ratio varies from 4:1 to 8:1 and the air-fuel mixture from 10:1 to 20:1.

Classification of Petrol Engines:

- Two Stroke cycle Petrol Engines.
- Four Stroke cycle petrol Engines.

Two Stroke Cycle Petrol Engine

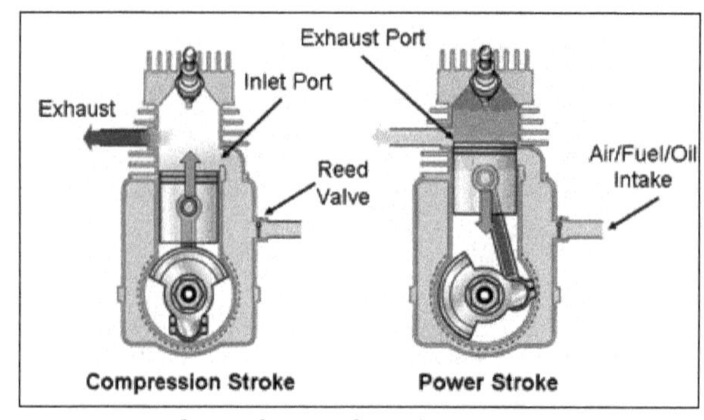

Two Stroke Cycle Petrol Engine - Construction.

Construction

A piston reciprocates inside the cylinder. It is connected to the crankshaft by means of connecting rod and crank. There are no valves in two stroke engines, instead of valves ports are cut on the cylinder walls. There are three ports, namely inlet, exhaust and transfer ports. The closing and opening of the ports are obtained by the movement of piston. The crown of piston is made in to a shape to perform this. A spark plug is also provided.

First Stroke: Upward Stroke of Piston

(a) Compression

The piston moves up from Bottom Dead Centre (BDC) to Top Dead Centre (TDC). Both transfer and exhaust ports are covered by the piston. Air fuel mixture which is transferred already into the engine cylinder is compressed by moving piston. The pressure and temperature increases at the end of compression.

(b) Ignition and Inductance

Piston almost reaches the top dead centre. The air fuel mixture inside the cylinder is ignited by means of an electric spark produced by a spark plug. At the same time, the inlet port is uncovered by the plane. Fresh air fuel mixture enters the crankcase through the inlet port.

Second Stroke: Downward Stroke of the Engine

(c) Expansion and Crankcase compression

The burning gases expand in the cylinder. The burning gases force the piston to move down. Thus, useful work is obtained. When the piston moves down, the air fuel mixture in the crankcase is partially compressed. This compression is known as Crankcase compression.

(d) Exhaust and Transfer

At the end of expansion, exhaust port is uncovered. Burnt gases escape to the atmosphere. Transfer port is also opened. The partially compressed air fuel mixture enters the cylinder through the transfer port. The crown of the piston is made of a deflected shape. So the fresh charge entering the cylinder is deflected upwards in the cylinder. Thus, the escape of fresh charge along with the exhaust gases is reduced.

Four Stroke Cycle Petrol Engines

Construction

A piston reciprocates inside the cylinder. The piston is connected to the crank shaft by means of a connecting rod and crank. The inlet and exhaust valves are mounted on the cylinder head. A spark is provided on the cylinder head. The fuel used is petrol.

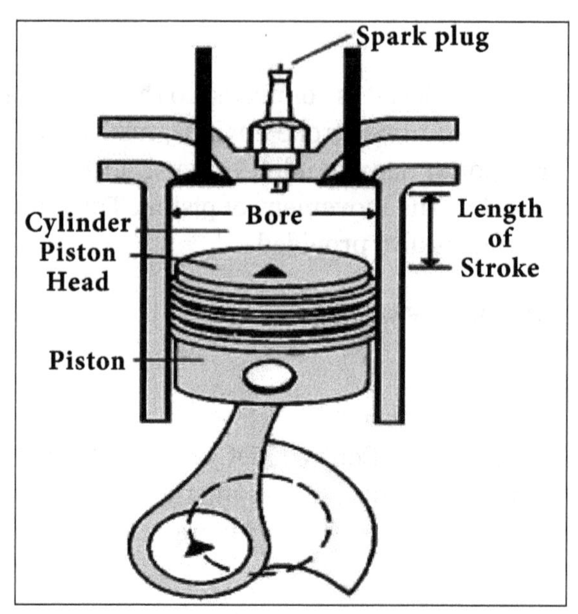

4 stroke petrol engine.

Working

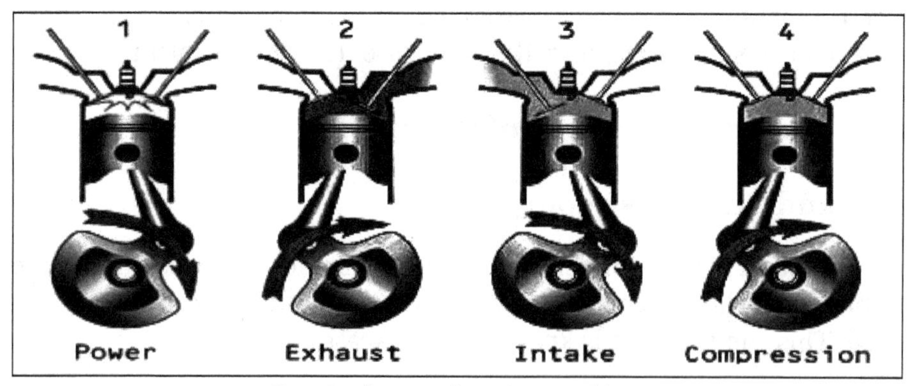

Four Stroke Petrol Engine-Working.

(a) Suction Stroke (First Stroke of the Engine)

- Piston moves down from TDC to BDC.
- Inlet valve is opened and the exhaust valve is closed.
- Pressure inside the cylinder is reduced below the atmospheric pressure.
- The mixture of air fuel is sucked into the cylinder through the inlet valve.

(b) Compression Stroke: (Second Stroke of the piston)

- Piston moves up from BDC to TDC.
- Both inlet and exhaust valves are closed.
- The air fuel mixture in the cylinder is compressed.

(c) Working or Power or Expansion Stroke: (Third Stroke of the Engine)

- The burning gases expand rapidly. They exert an impulse (thrust or force) on the piston. The piston is pushed from TDC to BDC.
- This movement of the piston is converted into rotary motion of the crankshaft through connecting rod.
- Both inlet and exhaust valves are closed.

(d) Exhaust Stroke (Fourth stroke of the piston)

- Piston moves upward from BDC.
- Exhaust valve is opened and the inlet valve is closed.
- The burnt gases are forced out to the atmosphere through the exhaust valve (Some of the burnt gases stay in the clearance volume of the cylinder).
- The exhaust valve closes shortly after TDC.
- The inlet valve opens slightly before TDC and the cylinder is ready to receive fresh charge to start a new cycle.
- Compression ratio varies from 5 to 8.
- The pressure at the end of compression is about 6 to 12 bar.
- The temperature at the end of the compression reaches 250° C to 350° C.

Four Stroke Diesel Engines

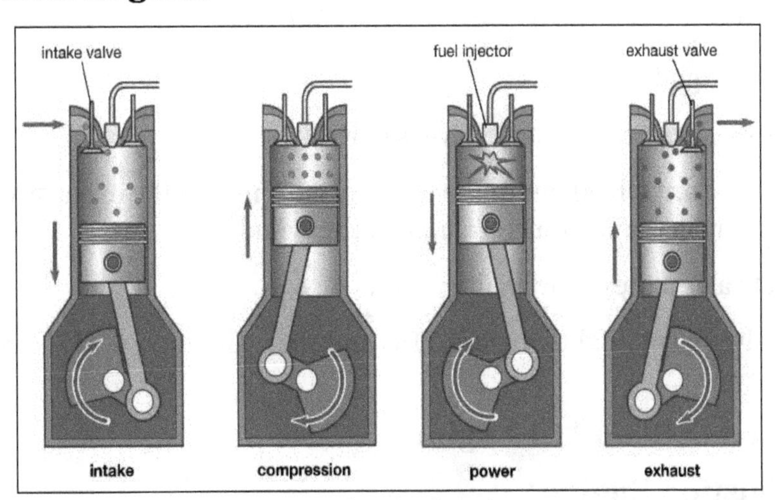

Four Stroke Diesel Engines.

Construction

A piston reciprocates inside the cylinder. The piston is connected to the crankshaft by means of a connecting rod and crank. The inlet and exhaust valves are mounted on the cylinder head. A fuel injector is provided on the cylinder head. The fuel used is diesel.

(a) Suction Stroke (First Stroke of the Piston)

- Piston moves from TDC to BDC.
- Inlet valve is opened and the exhaust valve is closed.
- The pressure inside the cylinder is reduced below the atmospheric pressure.
- Fresh air from the atmosphere is sucked into the engine cylinder through air cleaner and inlet valve.

(b) Compression stroke (Second Stroke of the Piston)

- Piston moves from BDC to TDC.
- Both inlet and exhaust valves are closed.
- The air is drawn during suction stroke is compressed to a high pressure and temperature.

(c) Working or Power or Expansion Stroke (Third Stroke of the Piston)

- The burning gases (products of combustion) expand rapidly.
- The burning gases push the piston move downward from TDC to BDC.
- This movement of piston is converted into rotary motion of the crank shaft through connecting rod.
- Both inlet and exhaust valves are closed.

(d) Exhaust Stroke (Fourth Stroke of the Piston)

- Piston moves from BDC to TDC.
- Exhaust valve is opened the inlet valve is closed.
- The burnt gases are forced out to the atmosphere through the exhaust valve. (Some of the burnt gases stay in the clearance volume of the cylinder)
- The exhaust valve closes shortly after TDC.
- The inlet valve opens slightly before TDC and the cylinder is ready to receive fresh air to start a new cycle.

2.2.1 Comparison of SI and CI Engines

External Combustion Engine

In these types of engines, combustion of fuel takes place outside the cylinder as in steam engines where the heat of combustion is employed to generate steam which is used to move a piston in a cylinder. Other examples include hot air engines, steam turbine and closed cycle gas turbine. These engines are used for driving locomotives, ships, generation of electric power etc.

Internal Combustion Engine

In these types of engines, combustion of the fuel with oxygen occurs within the cylinder of the engine. The internal combustion engines includes engines employing mixtures of combustible gases and air known as gas engines, those using lighter liquid fuel or spirit are known as petrol engines and those using heavier liquid fuels are known as oil compression ignition or diesel engines.

S.no:	SI Engine	Cl Engine
(i)	Basic Cycle: SI engine is operated by Otto cycle or constant volume cycle.	CI engine is operated by Diesel cycle or constant pressure cycle.
(ii)	Fuel Used: Fuel used for SI engine is petrol.	Fuel used for CI engine is diesel.
(iii)	Introduction of Fuel: In SI engine, the fuel is introduced to the cylinder along with air through the Inlet valve-during the suction stroke.	In Cl engine, the fuel is injected by the injector at the end of compression stroke.
(iv)	Ignition: In SI engines, the fuel- air mixture is ignited by a high-tension spark plug. Hence it is called as spark ignition engines.	In CI engines, the ignition of fuel air mixture takes place due to the high pressure and temperature of the air. Hence, they are known compression as Ignition Engines.
(vii)	Compression Ratio: Compression ratio for SI engine varies from 6 to 8	Compression ratio for CI engine varies from 12 to 18.
(viii)	Speed: These are used for high speed applications	These are used for low speed operations.

2.3 Valve Timing Diagram, Power Output and Efficiency Calculation

Typical Valve Timing Diagram for 4 Stroke Petrol Engine

The exact moment at which each of the valves open and close with reference to the position of piston and crank can be shown graphically in a diagram. This diagram is known as “Valve timing diagram”.

Theoretical Valve Timing Diagram

In theoretical valve timing diagram, inlet and exhaust valves open and close at both the dead centers. Similarly, all the processes are sharply completed at the TDC or BDC. Figure shows theoretical valve timing diagram for four strokes S.I. Engines.

Where,

- IVO - Inlet Valve Open.
- IVC - Inlet Valve Close.
- IS - Ignition Starts.
- EVO - Exhaust Valve Open.

- EVC - Exhaust Valve Close.
- TDC - Top Dead Center.
- BDC - Bottom Dead Center.

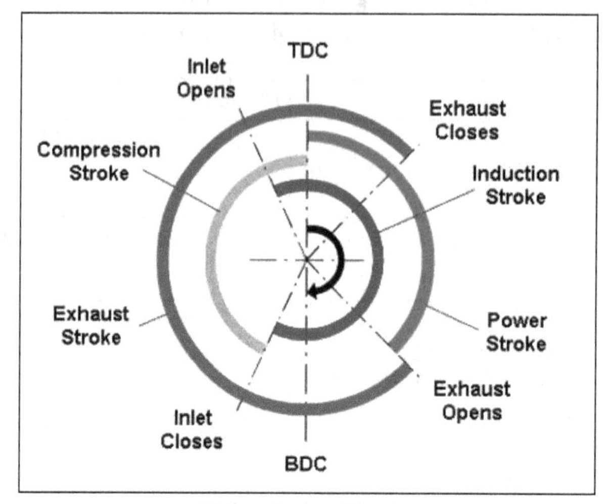

Valve timing Chart.

Actual Valve Timing Diagram

A figure shows actual valve timing diagram for four stroke S.I. engine. The inlet valve opens 10-30° before the TDC. The air-fuel mixture is sucked into the cylinder till the inlet Valve closes. The inlet valve closes 30-40° or even 60° after the BDC.

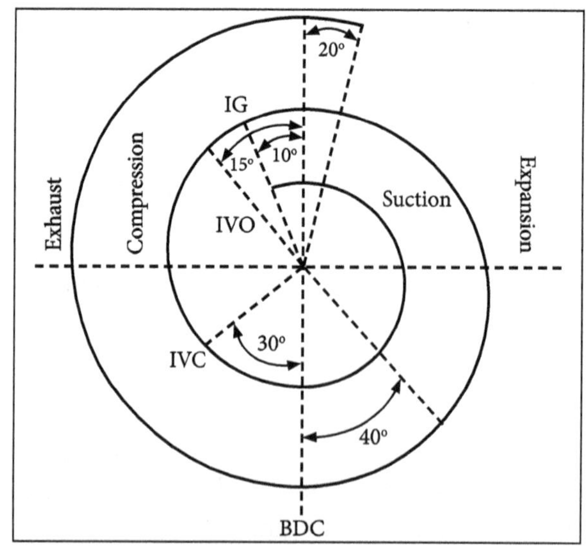

Actual valve timing diagram.

The charge is compressed till the spark occurs. The spark is produced 20-40° before the TDC. It gives sufficient time for the fuel to burn. The pressure and temperature increase. The burnt gases are expanded till the exhaust valve opens.

The exhaust valve opens 30-60° before the BDC. The exhaust gases are forced out from the cylinder till the exhaust valve closes. The exhaust valve closes 8-20° after the TDC. Before closing, again the inlet valve opens 10-30° before the TDC. The period between the IVO and EVC is known as valve overlap period. The angle between the inlet valve opening and exhaust valve closing is known as angle of valve overlap.

Port Timing Diagram for Two-Stroke Diesel Engine

1. Inlet Port

The Inlet Port opens at 35° to 50° prior to the TDC position which closes in equal amount after TDC position.

2. Exhaust Port

The exhaust port opens and closes at 35° -70° before and after BDC Position respectively.

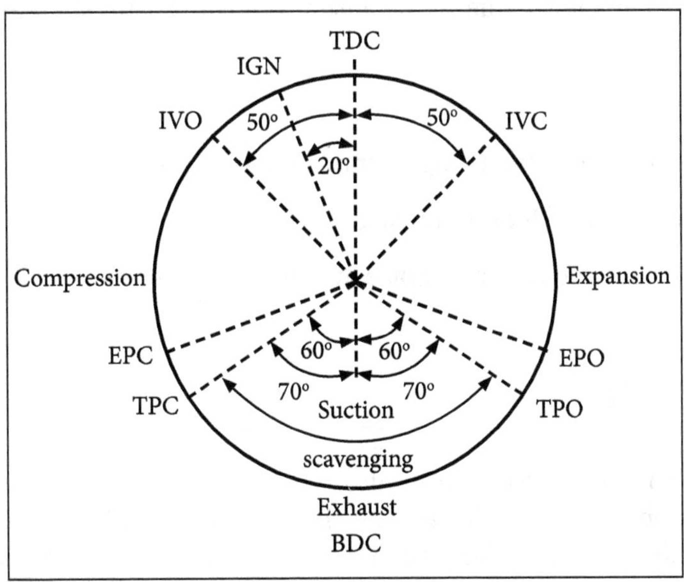

Port Timing Diagram.

Where,

- T.D.C - Top Dead Centre.
- I.P.O - Inlet Port Open.
- I.P.C - Inlet Port Close.
- T.P.O - Transfer Port Open.
- T.P.C - Transfer Port Close.
- E.P.O - Exhaust Port Open.
- E.P.C - Exhaust Port Close.
- B.P.C - Bottom Dead Centre.

3. Transfer Port

The transfer port opens at 35° to 60° in advance to the BDC position and closes at 35° to 60° after TDC position.

4. Ignition

Fuel injection valve opens at 10° to 15° before TDC position as the air requires some time to start ignition which closes at 15° to 20° after TDC position for better combustion. The scavenging period of petrol engines must not exceed above 70° whereas this period is large in case of diesel engines.

Advantages of two-stroke engines over four stroke engines:

- Construction is simple due to the absence of valves as the design of ports is much simpler and easy to manufacture.
- High mechanical efficiency due to the absence of cams, cam shaft and rockers etc., of the valve gear.

Disadvantages:

- Thermal efficiency is low due to high compression ratio.
- There is a great wear and tear of moving parts.
- Two stroke engines produce great noise during exhaust stroke.

2.4 Brayton Cycle

The Brayton cycle is also referred to as the Joule cycle or the gas turbine air cycle because all modern gas turbines work on this cycle. However, if the Brayton cycle is to be used for reciprocating piston engines, it requires two cylinders, one for compression and the other for expansion. Heat addition may be carried out separately in a heat exchanger or within the expander itself.

The pressure-volume and the corresponding temperature-entropy diagrams are shown in figure given below.

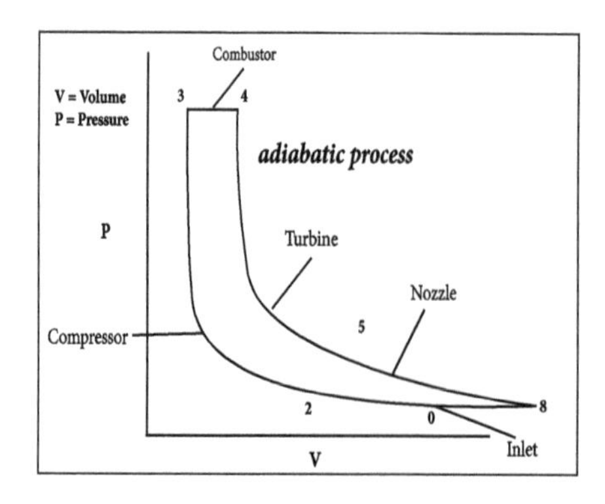

p - V diagram.

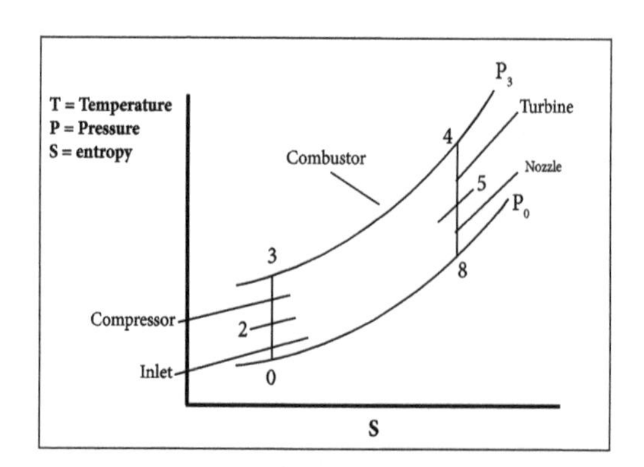

T - s diagram.

As we have seen the p - V and T - s diagrams for the Brayton cycle, we can now work on a problem from it.

The cycle consists of an isentropic compression process, a constant pressure heat addition process, an isentropic expansion process and a constant pressure heat rejection process. Expansion is carried out till the pressure drops to the initial (atmospheric) value.

Heat supplied in the cycle, Q_s, is given by,

$$C_v(T_3 - T_2)$$

Heat rejected in the cycle, Q_s, is given by,

$$C_v(T_4 - T_1)$$

Hence the thermal efficiency of the cycle is given by,

$$\eta_{th} = 1 - \left(\frac{T_4 - T_1}{T_3 - T_2}\right)$$

$$= 1 - \frac{T_1}{T_2}\left\{\frac{\left(\frac{T4}{T1} - 1\right)}{\left(\frac{T_3}{T_2} - 1\right)}\right\}$$

Now,

$$\frac{T_2}{T_1} = \left(\frac{P_2}{P_1}\right)^{\frac{\gamma-1}{\gamma}} = \left(\frac{P_3}{P_4}\right)^{\frac{\gamma-1}{\gamma}} = \frac{T_3}{T_4}$$

And since $\frac{T_2}{T_1} = \frac{T_3}{T_4}$, we have $\frac{T_4}{T_1} = \frac{T_3}{T_2}$

Hence, substituting in Equation we get, assuming that r_p is the pressure ratio p_2/p_1.

$$\eta_{th} = 1 - \frac{T_1}{T_2}$$

$$= 1 - \frac{1}{\left(\frac{p_2}{p_1}\right)^{\frac{\gamma-1}{\gamma}}}$$

$$= 1 - \frac{1}{r_p^{\frac{\gamma-1}{\gamma}}}$$

This is numerically equal to the efficiency of the Otto cycle if we put,

$$\frac{T_1}{T_2} = \left(\frac{V_2}{V_1}\right)^{\gamma-1} = \left(\frac{1}{r}\right)^{\gamma-1}$$

$$\eta_{th} = 1 - \frac{1}{r^{\gamma-1}}$$

Where, r is the volumetric compression ratio.

Problems

1. A gas turbine works on an air standard Brayton cycle. The initial condition of the air is 25° C and 1 bar. The maximum pressure and temperature are limited to 3 bar and 650° C. Let us determine the following:

- Cycle efficiency.
- Heat supplied and rejected per kg of air.
- Work output.
- Exhaust temperature.

Solution:

Given data:

$p_1 = p_4 = 1 \text{ bar} = 100 \text{ kN/m}^2$

$T_1 = 25°C = 298 \text{ K}$

$p_2 = p_3 = 3 \text{ bar} = 300 \text{ kN/m}^2$

$T_3 = 650°C = 923 \text{ K}$

To find:

- Cycle efficiency.
- Heat supplied and rejected per kg of air.
- Work output.
- Exhaust temperature

Formula to be used:

$$T_2 = \left(\frac{P_2}{P_1}\right)^{\frac{\gamma-1}{\gamma}} \text{x } T_1$$

$$T_4 = \left(\frac{P_4}{P_3}\right)^{\frac{\gamma-1}{\gamma}} \text{x } T_3$$

$$W_C = C_P(T_2 - T_1)$$

$$Q_r = C_p(T_6 - T_1)$$

Consider the process 1 - 2 isentropic compression:

$$\frac{T_1}{T_2} = \left(\frac{P_2}{P_1}\right)^{\frac{\gamma-1}{\gamma}}$$

$$T_2 = \left(\frac{P_2}{P_1}\right)^{\frac{\gamma-1}{\gamma}} \text{x } T_1$$

$$= \left(\frac{300}{100}\right)^{\frac{1.4-1}{1.4}} \text{x } 298$$

$$T_2 = 407.88\text{K}$$

Consider the process 3 - 4 isentropic expansion:

$$\frac{T_4}{T_3} = \left(\frac{P_4}{P_3}\right)^{\frac{\gamma-1}{\gamma}}$$

$$T_4 = \left(\frac{P_4}{P_3}\right)^{\frac{\gamma-1}{\gamma}} \text{x } T_3$$

$$= \left(\frac{100}{300}\right)^{\frac{1.4-1}{1.4}} \text{x } 923$$

$$T_4 = 1263.34\text{K}$$

Work output,

$$W_C = C_P(T_2 - T_1)$$
$$= 1.005 \times (1263.34 - 298)$$
$$W_c = 970.16 \text{ kJ}$$

Heat supplied,

$$Q_s = C_P(T_4 - T_3)$$
$$= 1.005 \times (1263.34 - 923)$$
$$Q_s = 342.04 \text{ kJ/kg}$$

Heat rejected,

$$Q_r = Cp\,(T_6 - T_1)$$
$$= 1.005\,(407.881 - 298)$$
$$Q_R = 110.43\ kJ/kg$$

2. Brayton cycle with air, pressure ratio is 12. Air enters the compressor at 300 K and enters the turbine at 1000 K. Net power output is 90 MW. Let us determine the required mass flow rate for turbine and compressor isentropic efficiencies of a) 100 percent b) 80 percent. Use constant specific heats

$$R = 0.287\ kJ/kg\ K,\ c_p = 1.005\ kJ/kg\ K$$

$$c_v = 0.718\ kJ/kg\ K,\ k = 1.4$$

$$T_1 = 300\ K \text{ and } T_3 = 1000\ K$$

Solution:

Given:

$$R = 0.287\ kJ/kg\ K,\ C_p = 1.005\ kJ/kg\ K$$

$$c_v = 0.718\ kJ/kg\ K,\ k = 1.4$$

$$T_1 = 300\ K \text{ and } T_3 = 1000\ K$$

To find:

The required mass flow rate for turbine and compressor isentropic efficiencies of: a) 100 percent and b) 80 percent. Use constant specific heats.

Formula to be used:

$$\frac{T_2}{T_1} = \left(\frac{P_2}{P_1}\right)^{(k-1)/k}$$

$$\dot{W}_{net} = \dot{m}\,c_p\left(T_1 - T_2 + T_3 - T_4\right)$$

$$\dot{m} = \frac{\dot{W}_{net}}{c_p\left(T_1 - T_2 + T_3 - T_4\right)}$$

a) The ideal case, process 1- 2 is isentropic, so we can use the isentropic ratio to find T_2:

$$\frac{T_2}{T_1} = \left(\frac{P_2}{P_1}\right)^{(k-1)/k} \rightarrow T_2 = T_1\left(r_p\right)^{(k-1)/k} = 300(2)^{0.4/1.4}$$

$$= 610.8\ k$$

We can do the same for process 3- 4 to find T_4:

$$T_4 = T_3\left(\frac{1}{r_p}\right)^{(k-1)/k} = 1000\left(\frac{1}{12}\right)^{0.4/1.4}$$

$$= 491.66 \text{ k}$$

The net power output of the Brayton cycle is:

$$\dot{W}_{net} = \dot{m}c_p\left(T_1 - T_2 + T_3 - T_4\right)$$

$$\dot{m} = \frac{\dot{W}_{net}}{c_p\left(T_1 - T_2 + T_3 - T_4\right)} = \frac{90{,}000}{1.005\left(300 - 610.18 + 1000 - 491.66\right)}$$

$$= 451.9 \text{kg/s}$$

b) The temperatures that we solved above are for the ideal case (T_2 , T_4). The actual work for the cycle is:

$$\dot{W}_{net} = \dot{m}c_p\left(T_1 - T_2\right)/\eta_c + \eta_c\ \dot{m}c_p\left(T_3 - T_4\right) = \dot{m}c_p\left[\frac{T_1 - T_2}{\eta_r} + \eta_r\left(T_3 - T_4\right)\right]$$

$$\dot{m} = \frac{\dot{W}_{net}}{c_p\left[\left(T_1 - T_2\right)/\eta_c + \eta_r\left(T_3 - T_4\right)\right]}$$

$$= \frac{90{,}000}{1.005\left[\left(300 - 610.18\right)/0.8 + 0.8\left(1000 - 491.66\right)\right]}$$

$$= 4726.5 \text{ kg/s}$$

Gas turbine

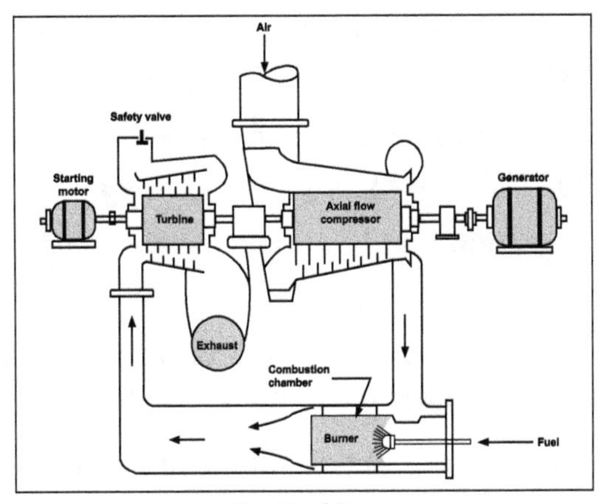

Gas Turbine.

A gas turbine is a combustion engine that can convert natural gas or other liquid fuels to mechanical energy. This energy then drives a generator that produces electrical energy. It is electrical energy that moves along power lines to homes and businesses.

Working

To generate electricity, the gas turbine heats a mixture of air and fuel at very high temperatures, causing the turbine blades to spin. The spinning turbine drives a generator that converts the energy into electricity. The gas turbine can be used in combination with a steam turbine in a combined-cycle power plant to create power extremely efficiently.

The gas turbine compresses air and mixes it with fuel that is then burned at extremely high temperatures, creating a hot gas. The hot air and fuel mixture moves through blades in the turbine, causing them to spin quickly.

Spinning blades turn the drive shaft. The fast spinning turbine blades rotate the turbine drive shaft. The spinning turbine is connected to the rod in a generator that turns a large magnet surrounded by coils of copper wire. Generator magnet causes electrons to move and creates electricity.

The fast revolving generator magnet creates a powerful magnetic field that lines up the electrons around the copper coils and then causes them to move. The movement of these electrons through a wire is electricity.

Merits of Gas Turbines Over I.C. Engines

- The mechanical efficiency of a gas turbine (96%) is quite high as compared with I.C. engine (85%) since the I.C. engine has a large number of sliding parts.
- A gas turbine does not require a flywheel as the torque on the shaft is continuous and uniform, whereas a flywheel is a must in case of an I.C. engine.
- The weight of gas turbine per H.P. developed is less than that of an I.C. engine.
- The gas turbine can be driven at very high speeds (40000 r.p.m.) whereas this is not possible with I.C. engines.
- The work developed by a gas turbine per kg of air is more as compared to an I.C. engine. This is due to the fact that gases can be expanded up to atmospheric pressure in case of a gas turbine whereas in an I.C. engine expansion up to atmospheric pressure is not possible.
- The components of the gas turbine can be made lighter since the pressures used in it are very low, say 5 bar compared with I.C. engine, say 60 bar.
- In the gas turbine the ignition and lubrication systems are much simpler as com-pared with I.C. engines.
- Cheaper fuels such as paraffin type, residue oils or powdered coal can be used whereas special grade fuels are employed in petrol engine to check knocking or pinking.
- The exhaust from gas turbine is less polluting comparatively since excess air is used for combustion.

- Because of low specific weight the gas turbines are particularly suitable for use in air-craft's.

Demerits of Gas Turbines

- The thermal efficiency of a simple turbine cycle is low (16 to 20%) as compared with I.C. engines (25 to 30%).
- With wide operating speeds the fuel control is comparatively difficult.
- Due to higher operating speeds of the turbine, it is imperative to have a speed reduction device.
- It is difficult to start a gas turbine as compared to an I.C. engine.
- The gas turbine blades need a special cooling system.
- One of the main demerits of a gas turbine is its very poor thermal efficiency at part loads, as the quantity of air remains same irrespective of load and output is reduced by reducing the quantity of fuel supplied.
- Owing to the use of nickel-chromium alloy, the manufacture of the blades is difficult and costly.
- For the same output the gas turbine produces five times exhaust gases than I.C. engine.
- Because of prevalence of high temperature (1000 K for blades and 2600 K for combustion chamber) and centrifugal force the life of the combustion chamber and blades is short/ small.

2.5 Jet Engines

The Brayton cycle is the basic cycle for jet engines. In the ideal jet propulsion cycle, the gases are not expanded to the ambient pressure as in the case of ideal Brayton cycle. Instead the gases are expanded to such a pressure that the power developed by the turbine is just sufficient to drive the compressor and the auxiliary equipment.

The net work output of a jet propulsion cycle is zero. The high pressure in the exhaust from the turbine of a jet engine is utilized for further expansion in a nozzle that results in the thrust to propel the aircraft.

The thrust or propulsive force of a jet engine is the difference between the momentum of the air entering the engine and that of the high velocity exhaust gases leaving the engine.

Thus, Thrust = Momentum at exit - Momentum at entry

Or

$$F = m'v_{exit} - m'v_{inlet} = m'(v_{exit} - v_{inlet})\,N$$

Where,

F -Thrust

Vexit- Velocity at exit of the exhaust gases. m/s

Vinlet -Velocity of air at inlet. m/s.

The velocities are reckoned relative to the aircraft. Since the fuel added is less than 2% of the air flow very link error is incurred by neglecting the fuel mass.

It should be noted that the pressure at entry and exit is both equal to the atmospheric pressure.

The propulsive power W_p, is defined as given,

Propulsive power = propulsive force x distance through which the force acts in unit time.

Or

$$W_p = F \times v_{aircraft}$$
$$= m'(v_{exit} - v_{inlet}) v_{aircraft}$$

If the heat input to the engine is Q_{in}', we can define the propulsive efficiency. η_p is given by,

$$\eta_p = \frac{\text{Propulsive power}}{\text{Energy input}} = \frac{W_p}{Q'_{in}}$$

Jet Engines May Be Classified into:

- Ramjets.
- Pulse jets.
- Turbojets.
- Turboprops.
- Rocket engines.

2.6 Reciprocating Air Compressors

The reciprocating Compressor has piston, cylinder, connecting rod, inlet valve, crank, exit valve, piston pin, crank pin and crank shaft. Inlet valve and exit valves may be of spring loaded type which get opened and closed due to pressure differential across them.

It consider piston to be at top dead centre (TDC) and move towards bottom dead centre (BDC). Due to this piston movement from TDC to BDC suction pressure is created causing opening of inlet valve. With this opening of inlet valve and suction pressure the atmospheric air enters the cylinder.

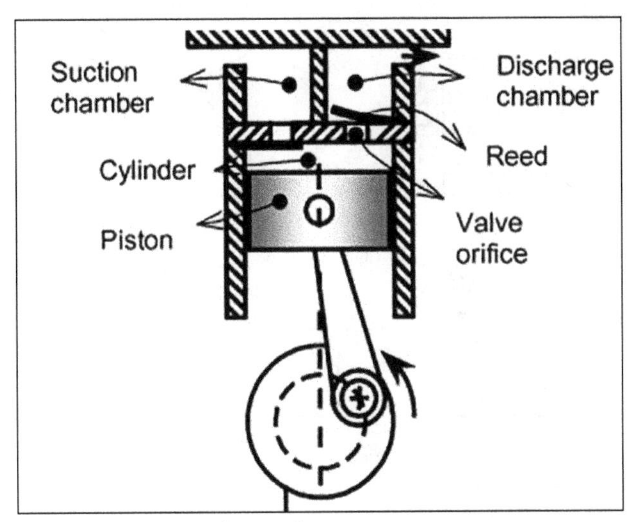

Reciprocating compressor.

Air gets into cylinder during this stroke and is subsequently compressed in next stroke with both inlet valve and exit valve closed. Each inlet valve and exit valves are of plate type and spring loaded so as to operate automatically as and when sufficient pressure difference is available to cause deflection in spring of valve plates to open them.

After piston reaching BDC it reverses its motion and compresses the air inducted in previous stroke. Compression is continued till the pressure of air inside becomes sufficient to cause deflection in exit valve. All the moment when exit valve plate gets lifted the exhaust of compressed air takes place. This piston again reaches TDC from where downward piston movement is again accompanied by suction.

In order to counter for the heating of piston-cylinder arrangement during compression the provision of cooling the cylinder is there in the form of cooling jackets in the body. Reciprocating compressor described above has suction, compression and discharge as three prominent processes getting completed in two strokes of piston or one revolution of crank shaft.

Applications

The reciprocating compressor generally seen where there is requirement of high pressure and low flow(or discontinuous flow up to 30 bars).Mostly where the air is used for hand-tools, cleaning dust, small paint jobs, commercial uses etc.

Single Acting Compressor

In a single stage compressor, the compression of air from the initial pressure to final pressure is carried out in one cylinder only. A schematic diagram of a single stage, single acting compressor is shown in figure. It consists of a cylinder, piston, connecting rod, crank, inlet and discharge values. When the piston moves downward i.e., during Suction stroke, the pressure of air inside the cylinder falls below the atmospheric pressure.

The inlet value opens and the air from atmospheric is sucked and the cylinder until the piston reaches the bottom dead center.

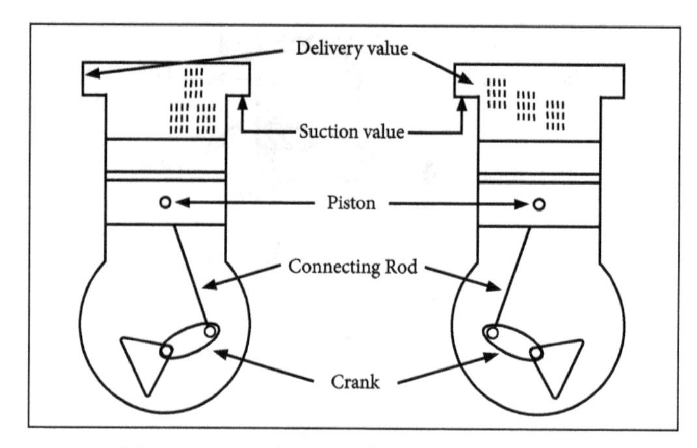

(a) Suction stroke. (b) Delivery stroke.

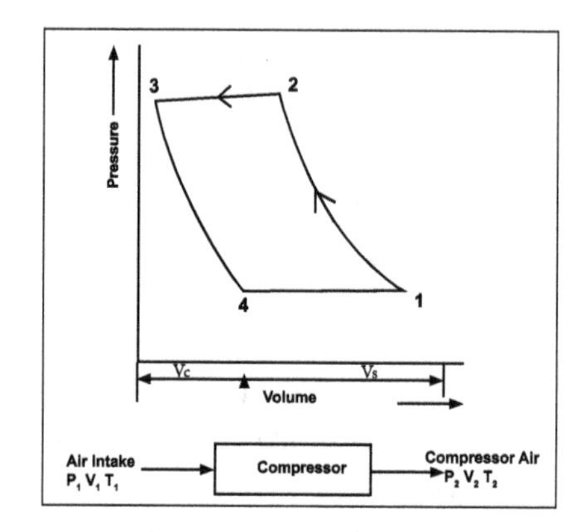

During this stroke delivery valve remains closed when the piston moves upwards both valves are closed. So the pressure inside the cylinder goes on increasing till it reaches required discharge pressure. At this stage, the discharge valve opens and the compressed air is delivered through this value. Thus the cycle is repeated.

Double Acting Compressor

The double stage or two stage reciprocating air compressors consists of two cylinders. One is called low pressure cylinder and another is known as high pressure cylinder.

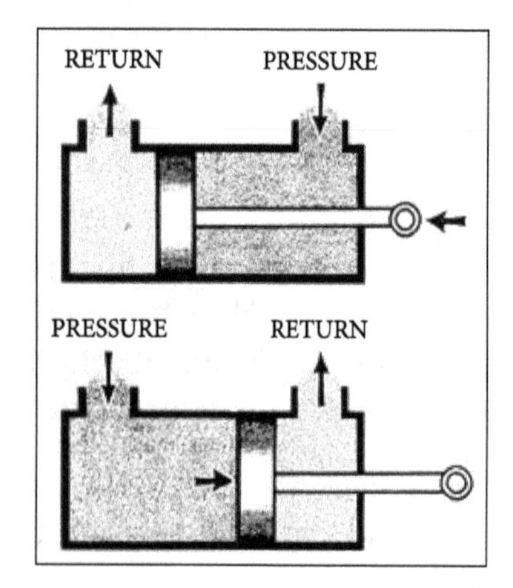

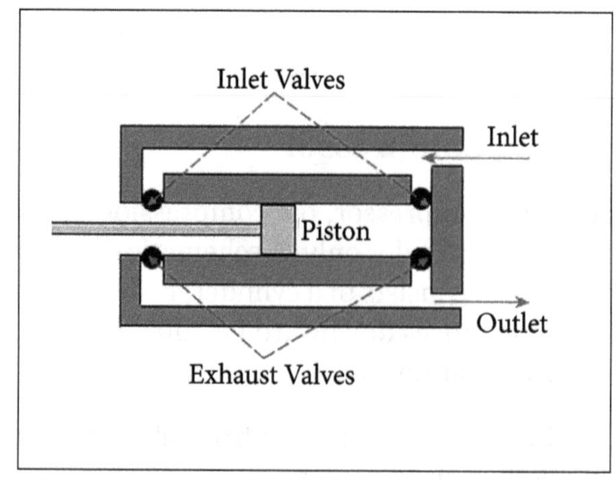

Double acting compressor.

When piston in low pressure cylinder is at its outer dead center (ODC) the weight of air inside cylinder is zero, as piston moves towards inner dead center (IDC) pressure falls below atmospheric pressure and suction valves opens due to pressure difference.

The fresh air is drawn inside the low pressure cylinder through air suction filter. This air is further compressed by piston and pressure inside & outside the cylinder is equal, at this point suction valves closed.

As piston moves towards ODC compression of air took place and when the pressure of air is in range of 1.5 kg/centimeter square to 2.5 kg/centimeter square delivery valves opens and this compressed air is then entered into high pressure cylinder through inter cooler. This called as low pressure compression.

If suction and discharge stroke took place on both side of piston then it is called Double Acting Low pressure compression.

The suction valves of high pressure cylinder opens when air pressure in high pressure side is below to the receiver pressure and air from low pressure cylinder drawn into high pressure cylinder.

As piston moves towards the ODC, from the beginning stage air is further compressed. When air pressure from low pressure cylinder and inside the high pressure cylinder is equal, suction valves closed.

Now air is further compressed by piston until the pressure in the High Pressure Cylinder exceeds that in the receiver and discharge valves opens. This desired high pressure air is then delivered to receiver.

Same procedure is repeated in every cycle of operation. If suction and discharge stroke took place on both side of piston then it is called double acting high pressure compression. In double stage reciprocating air compressor air pressure can be developed in range of 5.5 kg/centimeter square to 35 kg/centimeter square.

Normally where we required air pressure above 7.0 kg/cm^3 and delivery of air above 100 cubic feet/min. this double stage reciprocating air compressor is used. This is most common model used in various engineering plants, if we required air pressure above 35 kg/cm^3.

2.6.1 The Reciprocating Cycle Neglecting and Considering Clearance Volume

Theoretical Indicator Diagram Without Clearance and Compression

Clearance: The short distance between the piston and the cylinder cover is called linear piston clearance (L_c). The volume of space between the piston and the cylinder cover, when the piston is at the end of the stroke, plus the volume of the port leading to this space is called volumetric clearance (V_c). It is usually expressed as percentage of swept volume.

$$V_c = V_p + \frac{\pi}{4} D^2 . L_c \qquad ...(i)$$

Where, V_p = Port volume.

If V_p is neglected,

Then

$$V_c = \frac{\pi}{4} D^2 . L_c \qquad \text{...(ii)}$$

Piston clearance serves the following purposes:

- It checks the piston striking cylinder head.
- It allows small quantity of water to be collected due to steam condensation without excessive rise of pressure.
- It provides cushioning effect to the fast moving piston when it changes its direction of motion.

The work done per cycle by an engine is equal to the area of the p-V diagram shown in figure (a).

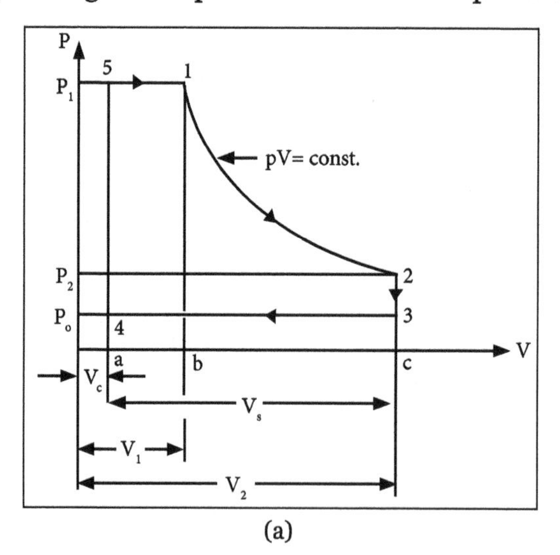

(a)

∴ Work done,

W = Area of the theoretical indicator diagram

$$= P_1V_1 + P_1V_1 \log_e V_2 / V_1 - P_bV_2$$

$$\text{Theoretical m.e.p} = \frac{\text{Theoretical work done}}{\text{Stroke volume}}$$

$$= \frac{p_1V_1 + p_1V_1 \log_e V_2 / V_1 - p_bV_2}{V_2} = \frac{p_1V_1}{V_2}(1 + \log_e V_2 / V_1) - p_b$$

The expansion ratio, r, is defined as $\frac{V_2}{V_1}$,

Therefore,

Theoretical/Hypothetical m.e.p,

$$p_m = \frac{p_1}{r}(1+\log_e r) - p_b \qquad \text{...(iii)}$$

Note:

- Cut off ratio $= \frac{V_1}{V_2} = \frac{1}{r}$, "Cut-off ratio is the ratio of volume between the points of admission and cut-off to the swept volume".
- A hypothetical steam consumption can be estimated using the specific volume, v, of the inlet steam and the volume induced per cycle, V_1, i.e., Steam consumption/cycle = $\frac{V_1}{v}$ kg.

Theoretical Indicator Diagram with Clearance

Work done,

$$W = \text{Area of indicator diagram} = \text{Area '512345' [Refer figure (b)]}$$

$$= \text{Area '51ba5'} + \text{Area '12cb1'} - \text{Area 43ca4}$$

$$= P_1(V_1 - V_c) + P_1V_1\log_e V_2/V_1 - P_b(V_2 - V_c)$$

$$[V_c = V_5 = V_4 = \text{Clearance volume}]$$

$$= P_1(V_1 - V_c) + P_1V_1\log_e V_2/V_1 - P_bV_s \qquad \text{...(iv)}$$

$$[\because V_2 - V_c = V_s = \text{Swept volume}]$$

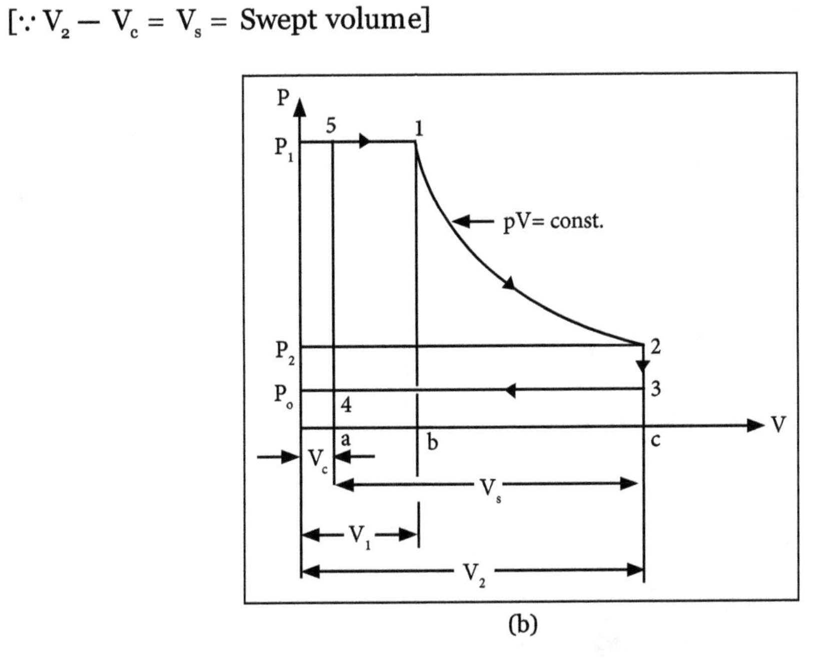

(b)

Let c = Ratio of clearance volume to swept volume

$$= V_c / V_s \qquad ...(v)$$

Also, $V_s = V_2 - V_c$ ($\because V_2 = V_s + V_c$)

Cut-off ratio,

$$\frac{1}{r} = \frac{V_1 - V_C}{V_S} \qquad \left(\because V_1 = \frac{V_S}{r} + V_c\right)$$

Inserting the value of V_c from equation (v), we get,

$$V_1 = \frac{V_S}{r} + c \cdot V_s$$

Or,

$$V_1 = V_s\left(c + \frac{1}{r}\right) \qquad ...(vi)$$

Inserting the values of V_c, V_2 and V_1 in equation (iv), we get,

$$W = p_1\left[V_s\left(c + \frac{1}{r}\right) - cV_s\right] + p_1\left(c + \frac{1}{r}\right)V_s \log_e\left[\frac{V_S(1+c)}{V_s\left(c + \frac{1}{r}\right)}\right] - p_b V_b$$

$$= p_1 V_s\left[c + \frac{1}{r} - c\right] + p_1 V_s\left(c + \frac{1}{r}\right)\log_e\left[\frac{1+c}{c + \frac{1}{r}}\right] - p_b V_b$$

$$= \frac{p_1 V_s}{r} + p_1 V_s\left(c + \frac{1}{r}\right)\log_e\left[\frac{1+c}{c + \frac{1}{r}}\right] - p_b V_b \qquad ...(vii)$$

$$p_m = \frac{\text{work done}}{\text{stroke volume}} = \frac{\frac{p_1 V_s}{r} + p_1 V_s\left(c + \frac{1}{r}\right)\log_e\left[\frac{1+c}{c + \frac{1}{r}}\right] - p_b V_b}{V_S}$$

$$\frac{p_1}{r} = p_1\left(c + \frac{1}{r}\right)\log_e\left[\frac{1+c}{c + \frac{1}{r}}\right] - p_b$$

$$p_m = p_1\left[\frac{1}{r}+\left(c+\frac{1}{r}\right)\log_e\left(\frac{1+c}{c+\frac{1}{r}}\right)\right]-p_b$$
...(viii)

Theoretical Indicator Diagram with Clearance and Compression

Work done,

$$W = \text{Area of the indicator diagram } (\text{Refer figure}(c))$$
$$= \text{Area '6123456'}$$
$$= \text{Area '61da6'} + \text{Area '12ed1'} - \text{Area 34be3} - \text{Area '45ab4'}$$

Following the same way as discussed in the case 2, we get the following expression for mean effective pressure.

$$p_m = p_1\left[\frac{1}{r}+\left(c+\frac{1}{r}\right)\log_e\left(\frac{c+1}{c+\frac{1}{r}}\right)\right]-p_b\left[(1-\alpha)+(c+\alpha)\log_e\left(\frac{\alpha+c}{c}\right)\right]$$
...(ix)

Where,

$$c=\frac{V_C}{V_S}, \frac{1}{r}=\frac{V1-V_C}{V_S}$$

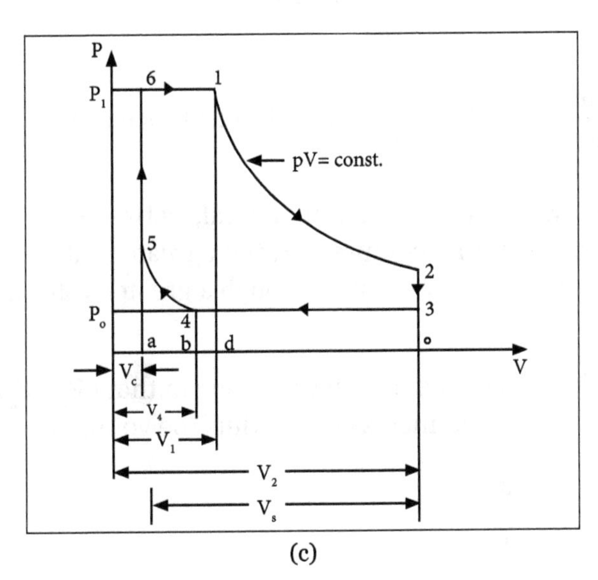

(c)

And α =ratio of the volume between points of compression and admission to the swept volume V_s

$$=\frac{V_4-V_C}{V_S}=\frac{V_4}{V_S}-c$$

2.6.2 Volumetric Efficiency and its Effect On Compressor Performance

Volumetric Efficiency of A Reciprocating Compressor

$$\eta_{vol} = \frac{\text{Actual capacity of the compressor}}{\text{Piston displacement of the compressor}} = \frac{V_a}{V_s}$$

$$\eta_{vol} = 1 + C - C\left(\frac{P_2}{P_1}\right)^{\frac{1}{\eta}}$$

$$\frac{P_2}{P_1} = \text{Pressure ratio}$$

$$C = \frac{V_c}{V_S} = \text{Clearance ratio}$$

Derivation for Volumetric Efficiency of an Air Compressor

Volumetric Efficiency

The volumetric efficiency of a compressor is the ratio of free air delivered to the displacement of the compressor. It is also the ratio of effective swept volume to the swept volume. i.e.,

$$\text{Volume efficiency} = \frac{\text{Effective swept volume}}{\text{swept volume}} = \frac{V_1 - V_4}{V_1 - V_3}$$

Because of presence of clearance volume, volumetric efficiency is always less than unity, as a percentage, it usually varies from 60% to 85%.

The ratio $\frac{\text{Clearance volume}}{\text{swept volume}} = \frac{V_3}{V_1 - V_3} = \frac{V_1}{V_3} = k$ is the clearance ratio.

As a percentage, this ratio will have a value, in general, between 4% and 10%. The greater the pressure ratio through a reciprocating compressor, then greater will be the effect of the clearance volume since the clearance air will now expand through a greater volume before intake conditions are reached.

The cylinder size and stroke being fixed, however will mean that $(V_1 - V_4)$, the effective swept volume, will reduce as the pressure ratio increases and thus the volumetric efficiency reduces.

$$\text{Volume efficiency, } \eta_{volume} = \frac{V_1 - V_4}{V_1 - V_3}$$

$$= \frac{(V_1 - V_3) - (V_3 - V_4)}{(V_1 - V_3)} = 1 + \frac{V_3}{V_1 - V_3} - \frac{V_4}{V_1 - V_3}$$

$$= 1 + \frac{V_3}{V_1 - V_3} - \frac{V_4}{V_1 - V_3} = 1 + \frac{V_3}{V_1 - V_3} - \frac{V_3}{V_1 - V_3} \cdot \frac{V_4}{V_3}$$

$$= 1 + k - k \cdot \frac{V_4}{V_3}$$

$$\left| \begin{array}{l} p_3 V_3^n = p_4 V_4^n \\ \dfrac{V_4}{V_3} = \left(\dfrac{p_3}{p_4} \right)^{1/n} \end{array} \right|$$

$$= 1 + k - k \cdot \left(\frac{p_3}{p_4} \right)^{1/n}$$

$$\text{or, } \eta_{vol} = 1 + k - k \left(\frac{p_2}{p_1} \right)^{1/n}$$

$$(\because p_3 = p_2, p_4 = p_1)$$

$$\text{or, } \eta_{vol} = 1 + k - k \left(\frac{V1}{V_2} \right)$$

The above equations are valid if the index of expansion and compression is same. However it may be noted that the clearance volumetric efficiency is dependent on only the index of expansion of the clearance volume from V_3 to V_4. Thus, if the index of compression = n_c and index of expansion = n_e, the volumetric efficiency is given by,

$$\eta_{vol} = 1 + k - k \left(\frac{p_3}{p_1} \right)^{1/n}$$

$$= 1 + k - k \left(\frac{p_2}{p_1} \right)^{1/n}$$

$$= 1 + k - k \left(\frac{V_4}{v_3} \right)$$

In this case volumetric efficiency is $= 1 + k - k \left(\frac{}{} \right)$.

Isothermal Efficiency

The isothermal compression represents the ideal minimum work input for a compression process (where cooling is feasible).

We can therefore define compressor efficiency as:

$$\eta_{iso} = \frac{\text{Isothermal work per cycle}}{\text{Actual work per cycle}}$$

If we recalculate the work input assuming isothermal compression [$W_{12} = p_1V_1 \ln(p_1/p_2)$ etc.] it is found that:

$$\eta_{iso} = \frac{\ln r_p}{\frac{n}{n-1}\left\{r_p^{\frac{n-1}{n}} - 1\right\}}$$

This efficiency is known as the isothermal efficiency.

Isentropic Efficiency of Reciprocating Compressor

$$\eta_{vol} = \frac{\text{Volume of free air taken per cycle}}{\text{Stroke volume of the cylinder}}$$

An adiabatic process is one in which no heat is added or removed from the system. Adiabatic compression is expressed by,

$$P_1V_1^k = P_2V_2^k \ (1)$$

Where, $k = C_p/C_v$ = ratio of specific heats, dimensionless.

Although compressors are designed to remove as much heat as possible, some heat gain is inevitable. Nevertheless, the adiabatic compression cycle is rather closely approached by most positive displacement compressors and is generally the base to which they are referred.

p -V diagram of a two stage reciprocating air compressor

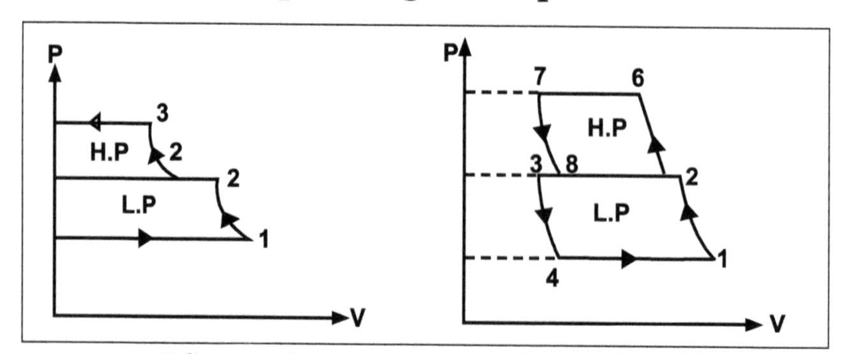

p -V diagram of a two stage reciprocating air compressor.

Without considering clearance volume

- L.P - Low Pressure Cylinder.
- H.P - High Pressure Cylinder.

2.6.3 Limitations of Single Stage Compression

The pressure ratio for a single-stage reciprocating air-compressor is limited to seven increases in pressure ratio in a single-stage reciprocating air compressor causes the following undesirable effects:

- Greater expansion of clearance air in the cylinder and as a consequence, it decreases effective suction volume (V_1 - V_4) and therefore, there is a decrease in fresh air induction.
- With high delivery pressure, the delivery temperature increases. It increases specific volume of air in the cylinder, thus more compression work is required.
- Further, for high pressure ratio, the cylinder size would have to be large, strong and heavy working pans of the compressor will be needed. It will increase balancing problem and high torque fluctuation will require a heavier flywheel installation.

Problems

1. Let us estimate the volumetric efficiency and power consumption of a single stage single acting reciprocating compressor, given the following data: Cylinder diameter: 30 cm, Stroke: 22 cm, Clearance ratio: 0.03, Delivery pressure: 8 bar, Suction pressure: 1 bar, Speed: 400 rpm, Compression and expansion follows $pv^{1.3}$ = constant.

Solution:

Given Data:

D = 30 cm = 0.3 m

L = 22 cm = 0.22 m

C = 0.03

P_2 = 8 bar = 800 kpa

P_1 = 1 bar = 100 kpa

N = 400 rpm

$Pv^{1.3} = C$

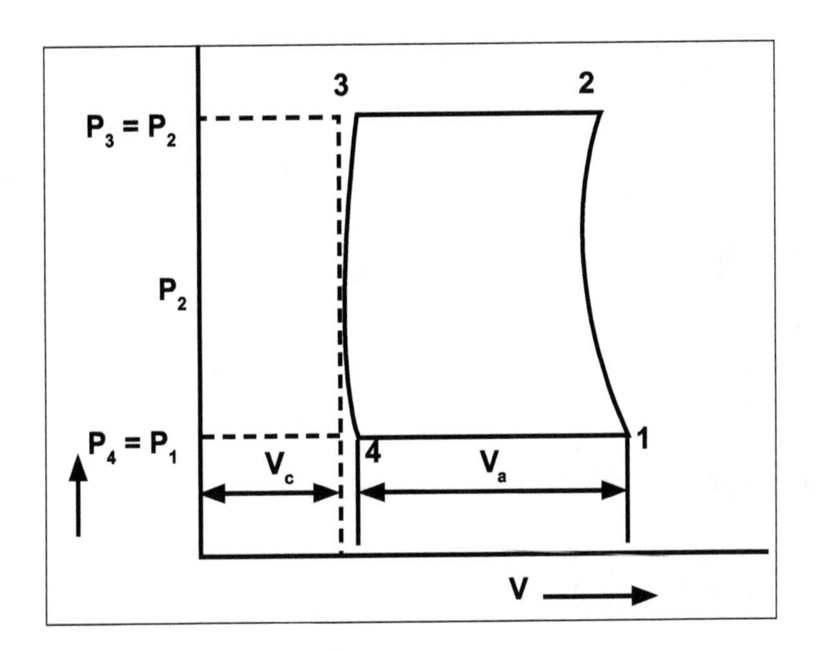

To find:

- Volumetric efficiency η_{vol}.
- Power consumption P.

Formula to be used

$$\eta_{vol} = 1 - \frac{V_C}{V_S}\left[\frac{V_4}{V_C} - 1\right] x\ 100$$

$$\eta_{vol} = 1 + C - C\left[\left(\frac{P_2}{P_1}\right)^{\frac{1}{n}}\right]$$

$$W = \frac{n}{n-1} P_1 V_a \left[\left(\frac{P_2}{P_1}\right)^{\frac{n-1}{n}} - 1\right]$$

$$\text{Clearance ratio} = \frac{V_C}{V_S}$$

Volumetric Efficiency

$$\eta_{vol} = \text{Volume of free air taken per cycle and stock volume of the cylinder}.$$

$$\eta_{vol} = 1 - \frac{V_C}{V_S}\left[\frac{V_4}{V_C} - 1\right] x\ 100$$

(Or)

$$\eta_{vol} = 1 + C - C\left[\left(\frac{P_2}{P_1}\right)^{\frac{1}{n}}\right]$$

$$= 1 + 0.03 - 0.03\left[\left(\frac{8}{1}\right)^{\frac{1}{13}}\right]$$

$$= 0.88143 \times 100$$

$$n_{vol} = 88.147\%$$

Work done by the compressor with clearance volume,

$$W = \frac{n}{n-1} P_1 V_a \left[\left(\frac{P_2}{P_1}\right)\frac{n-1}{n} - 1\right]$$

$\therefore V_a = V_1 - V_4$

$V_1 = V_c + V_s$

$$\text{Clearance ratio} = \frac{V_C}{V_S}$$

$$V_s = \frac{\pi}{4} D^2 C = \frac{\pi}{4} \times (0.30)^2 \times 0.22$$

$$= 0.01555 m^3$$

$$C = \frac{V_c}{V_S} \Rightarrow V_c = C \times V_3$$

$= 0.03 \text{ x } 0.01555$

$V_c = 0.000466 \text{ m}^3$

Initial Compression of the spring,

$$\delta = \frac{S_1}{S} = \frac{426.36}{3.716 \text{ x } 10^3}$$

$\delta = 0.1147 \text{ mm}$

$\delta = 114.7 \text{ mm}$

Equilibrium speed N_2 = 5% greater than N_1

$N_2 = N + 0.05 N$

$= 1.05 N$

$N_2 = 1.05 \times 360$

$N_2 = 378 \text{ rpm}$

2. A single stage double acting air compressor delivers 15 m³/min of air measured at 1.013 bar, 27°C. The air is delivered at 7 bar. The conditions at the end of suction stroke air pressure 0.98 bar and temperature 35°C. The clearance volume is 4% of stroke volume, the L/D ratio is 1.3 and the compressor runs at 300 rpm. Let us calculate the volumetric efficiency, cylinder dimensions and isothermal efficiency of the compressor, taking index of expansion and compression as 1.3 and R = 0.287 kJ/kg.K.

Solution:

Given:

$V_o = 15 m^3/\text{min} = 0.25 \text{ m}^3/\text{s}$

$P_o = 1.013 \text{ bar} = 101.3 \text{kPa}$

$T_o = 27^o c + 273^o c = 300 k$

$P_2 = 7 \text{bar} = 700 \text{kPa}$

$P_1 = 0.98 \text{ bar} = 98 \text{kPa}$

$V_C = 4\% V_c = 0.04 V_3$

$$\frac{V_c}{V_3} = 0.04$$

$T_1 = 35°C = 308 \text{ K}$
$N = 300 \text{ rpm}$

$$\frac{L}{D} = 1.3$$

$R = 0.287 \text{ KJ/Kg}$
$n = 1.3$

To Find:

- $\eta_{volumetric}$.
- L and D of cylinder.
- $\eta_{Isothermal}$.

Formula to be used:

$$\eta_{vol} = HC - C\left[\left(\frac{P_2}{P_1}\right)^{1/n}\right]$$

$$V_o = V_s \times \eta_v \times N$$

$$W = \frac{n}{n-1} P_1 V_a \left[\left(\frac{P_2}{P_1}\right)^{\frac{n-1}{n}} - 1\right]$$

$$\eta_{vol} = HC - C\left[\left(\frac{P_2}{P_1}\right)^{1/n}\right]$$

$$= 1 + \frac{V_c}{V_s} - \frac{V_c}{V_s}\left[\left(\frac{P_2}{P_1}\right)^{1/n}\right]$$

$$= 1 + 0.04 - 0.04\left[\left(\frac{700}{98}\right)^{1/13}\right]$$

$$= 85.84\ \%$$

$$V_o = V_s \times \eta_v \times N$$

$$15 = V_s \times 0.8584 \times 300$$

$$1\ (\text{Stroke volume})\ V_S = 0.0582 \text{ m}^3$$

We know that,

$$V_s = \frac{\pi}{4} \times D^2 \times L = 0.0582$$

$$\frac{\pi}{4} \times D^2 \times 1.3D = 0.0582$$

$$D = 0.3849m$$

$$\frac{L}{D} = 1.3$$

$$L = 1.3 \times 0.3849$$

$$L = 0.5 \text{ m}$$

Isothermal efficiency = Isothermal power / Indicated power.

Isothermal power $= mRT_1 in(P_2 / P_1)$

$$= P_1V_a(P_2 / P_1)$$

To find V_a

But $\frac{P_0V_0}{T_0} = \frac{P_1V_4}{T_1}$

$$\frac{101.3 \times 15 \times 60}{300} = \frac{98 \times V_a}{308}$$

$$V_a = 0.265 \text{ m}^3 / \text{sec}$$

$$\text{Isothermal Power} = 98 \times 0.265 \, In \left(\frac{700}{98}\right)$$

51.05 KW

$$W = \frac{n}{n-1} P_1V_a \left[\left(\frac{P_2}{P_1}\right)^{\frac{n-1}{n}} - 1\right]$$

$$= \frac{1.3}{0.3} \text{ x } 98 \text{ x } 0.265 \left[\left(\frac{700}{98}\right)^{0.3/1.3} - 1\right]$$

$= 64.61 \text{ KJ/sec}$

$\Rightarrow$ Indicated power $= 64.61$ kW

$$\text{Isothermal efficiency} = \frac{51.05}{64.61} = 79\%$$

Result:

- $\eta_{VOL} = 85.84\%$.
- $L = 0.5$ m; $D = 0.3849$ m.
- $\eta_{Iso} = 79\%$.

3. A two cylinder single acting air compressor is to deliver 16 kg of air per minute a 7 bar from suction conditions 1 bar and 15°C. Clearance may be taken as 4% of stroke volume and the index for both compression and re-expression as 1.3. Compressor is directly coupled to a four cylinder four stroke petrol engine which runs at 2000 rpm with a brake mean effective pressure of 5.5 bar. Assuming a stroke-bore ratio of 1.2 for both engine and compressor and a mechanical efficiency of 82% for compressor, let us calculate the required cylinder dimensions.

Solution:

Given:

$m = 16$ Kg/min

$P_1 = 1$ bar

$P_3 = 7$ bar

$P_{mb} = 5.5$bar

$T_1 = 15°C = 288$ k

$V_c = 4\% V_5$

$h = 1.3$

$N = 2000$ rpm.

Brake mean effective pressure = 5.5 bar.

Stroke-bor ratio $\frac{L}{D} = 1.2$

$\eta_{mech} = 82\%$

To find:

- Cylinder dimensions.
- Amount of air delivered per cylinder $= 16/2 = 8$ *kg*/min.

From gas equation,

$$P_1(V_1 - V_4) = mRT_1$$

$$V_1 - V_4 = \frac{mRT_1}{P_1} = \frac{8 \times 287 \times 288}{1 \text{ x } 10^5}$$

$$V_1 - V_4 = 6.61 \text{ m}^3/\text{min}$$

$$= \frac{6.61}{N} = \frac{6.61}{2000} = 0.003305 \text{ m}^3/\text{stroke}$$

∴ Compressor being single acting.

From expansion curve,

$$\frac{V_4}{V_3} = \left[\frac{P_3}{P_4}\right]^{1/n} = \left[\frac{7}{1}\right]^{1/1.3}$$

$$= 4.467$$
$$V_4 = 4.467 \, V_3$$
$$= 4.467 \times 0.04 V_s$$

$$= 0.1787 \, V_s$$
$$V_1 + V_3 + V_s$$

$$= 0.04 \, V_S + V_s$$
$$= 1.04 Vs$$

$$V_1 - V_4 = 1.04 \, V_s - 0.1787 \, V_s$$
$$V_1 - V_4 = 0.8613 \, V_s$$

But $V_1 - V_4$ is also equal to 0.003305

$$\therefore 0.8613 \, V_S = 0.003305$$

$$V_s = \frac{0.003305}{0.8613}$$

$$V_s = 0.003837 \text{ m}^3$$

If L_C = Length of stroke of compressor and

D_C =Diameter of the cylinder of the compressor, then

$$L_C = 1.2\ D$$

$$V_s = \frac{\pi}{4} D^2_c L_c$$

$$0.003837 = \frac{\pi}{4} D^2_c \times 1.2\ D_c$$

$$D^3_c = \frac{0.003837 \times 4}{\pi \times 1.2}$$

$$D_c = 0.1596 \text{ m or } 159.6 \text{ mm}$$

$$L_C = 159.6 \times 1.2$$

$$L_C = 191.5 \text{ mm}$$

Now indicated power of the compressor,

$$= \frac{n}{n-1} mRT_1 \left[\left(\frac{P_2}{P_1} \right)^{\frac{n-1}{n}} - 1 \right]$$

$$= \frac{1.3}{1.3-1} \times \frac{16}{60} \times 0.287 \times 288 \left[\left(\frac{7}{1} \right)^{\frac{0.3}{1.3}} - 1 \right]$$

$$= 54.1 \text{ KW}$$

Brake power of the engine.

$$= \frac{54.1}{\eta_{mech}} = \frac{54.1}{0.82}$$

$$= 65.97 \text{ KW}$$

$$\text{Now, } B_p = \frac{\eta_e P_{m\delta} \cdot L_e AN}{60}$$

$$65.97 = \frac{\eta_e P_{m\delta} \cdot L_e AN}{60}$$

If D_e = Diameter of the engine cylinder.

L_e = Length of the stroke of the engine = 1.2 D

n_e = Number of engine cylinder

Then, $65.97 = \frac{\eta_e P_{m\delta} \cdot L_e AN}{60}$

$$65.97 = \frac{4 \times \left(5.5 \times 10^5\right) \times 1.2 D_e \times \frac{\pi}{4} \times D_e^2 \times 2000}{60}$$

$$D_e^3 = \frac{65.97 \times 60 \text{ x } 10^3 \times 4}{4 \times 10^5 \times 5.5 \times 1.2 \times \pi \times 2000}$$

$D_e^3 = 0.0009545 \text{ m}^3$

$D_e = 98.4 \text{ mm}$

$L_e = 1.2 \times D_2$

$L_e = 1.2 \times 98.4$

$L_e = 118.1 \text{ mm}$

2.7 Multistage Compression and Inter Cooling

In multistage compressor, the compression of the air from Pressure to the final pressure is carried out in more than one cylinder. It is used to get high pressure air. In a compressor, when the compression ratio exceeds five, generally multi-stage compression is adopted.

A schematic arrangement for a two stage reciprocating air compressor is shown in diagram. It consists of a Low Pressure cylinder (L.P) and intercooler and a High Pressure cylinder (H.P). Fresh air is sucked from the atmosphere in the low pressure cylinder during its Suction Stroke at intake Pressure P_1 and Temperature T_1.

The air after compression in the L.P cylinder is delivered to the inter cooler at Pressure P_2 and Temperature T_2. In the inter cooler, the air is cooled at constant pressure by circulating cold water. The cooled air from the inter cooler is then taken to the high pressure cylinder. In the high pressure cylinder, air is further compressed to the final delivery Pressure (P_3) and discharged to the receiver.

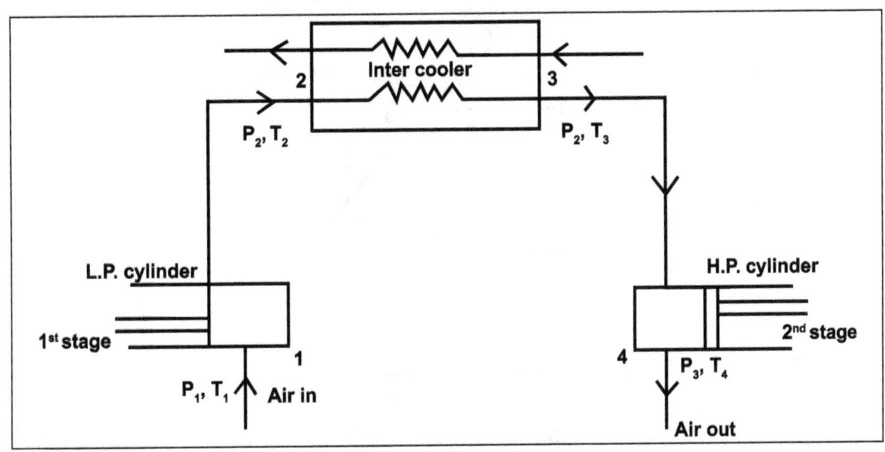

Multistage air compressor.

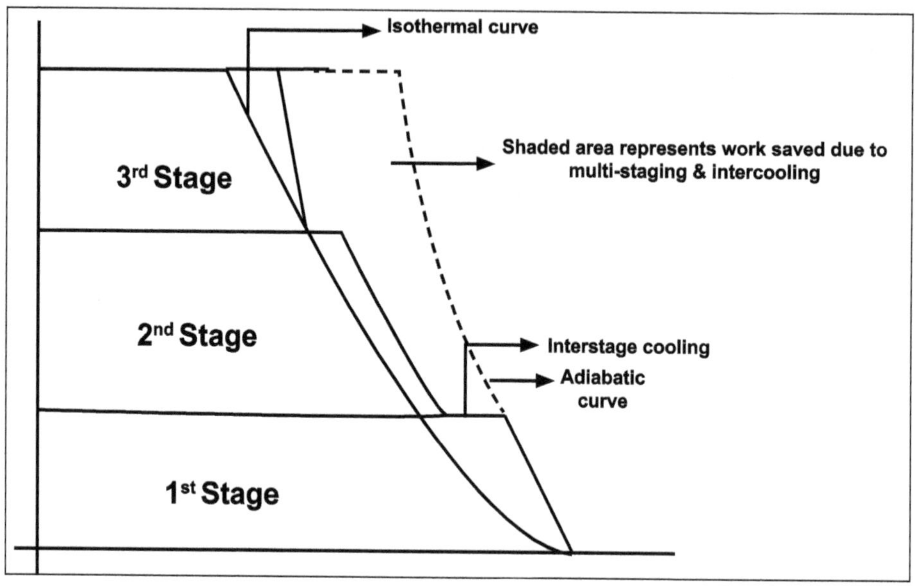

Isothermal curve.

Working of Multistage Compression with Inter Cooler

The following assumptions are made for two stage compressor with inter cooling:

- The effect of clearance is neglected.
- Compression is polytrophic ($PV^n = C$) for both the cylinders.
- There is no pressure drop in the inter cooler.
- Suction and delivery of air takes place at constant pressure.

The analysis is made in two cases.

Case (i): When the Inter cooling is Imperfect:

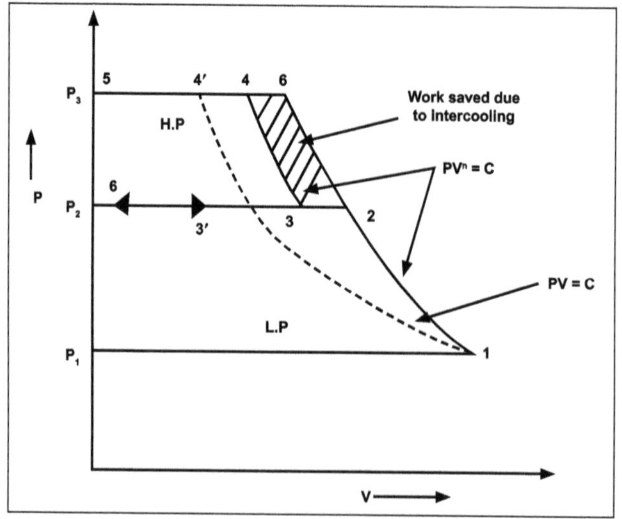

Inter-cooling is imperfect.

Two stage compression with inter cooling:

- P_1 - Pressure of air entering low pressure cylinder (LP)N/m².
- V_1 - Volume of air entering LP cylinder m³.
- P_2 - Pressure of air leaving the LP cylinder.
- V_2 - Volume of air leaving the LP cylinder.
- P_3 - Pressure of air leaving HP cylinder.
- V_3 - Volume of air leaving the HP cylinder.
- n - Index of compression for both the cylinder.

Work done in LP cylinder = Area (1-2-6-7-1)

Work done in HP cylinder = Area (3-4-5-6- 3)

Work saved due to inter cooling = Area (2-6-4-3- 2)

$$\text{Work required/Cycle in LP cylinder} = \frac{n}{n-1}(P_1V_1)\left[\left(\frac{P_2}{P_1}\right)^{\frac{n-1}{n}} - 1\right]$$

$$\text{Work required/Cycle in HP Cylinder} = \frac{n}{n-1}(P_3V_3)\left[\left(\frac{P_4}{P_3}\right)^{\frac{n-1}{n}} - 1\right]$$

$$\text{Total work required/Cycle, } W = \frac{n}{n-1}\left[(P_1V_1)\left[\left(\frac{P_2}{P_1}\right)^{\frac{n-1}{n}} - 1\right] + P_3V_3\left[\left(\frac{P_4}{P_3}\right)^{\frac{n-1}{n}}\right]\right]$$

$$W = \frac{n}{n-1}\left[(P_1V_1)\left[\left(\frac{P_2}{P_1}\right)^{\frac{n-1}{n}} - 1\right] + P_2V_2\left[\left(\frac{P_3}{P_2}\right)^{\frac{n-1}{n}}\right]\right]$$

Case (ii): When the inter cooling is perfect:

For perfect inter cooling,

$$P_1V_1 = P_2V_2$$

$$W = \frac{n}{n-1}(P_1V_1)\left[\left(\frac{P_2}{P_1}\right)^{\frac{n-1}{n}} + \left(\frac{P_4}{P_3}\right)^{\frac{n-1}{n}} - 2\right]$$

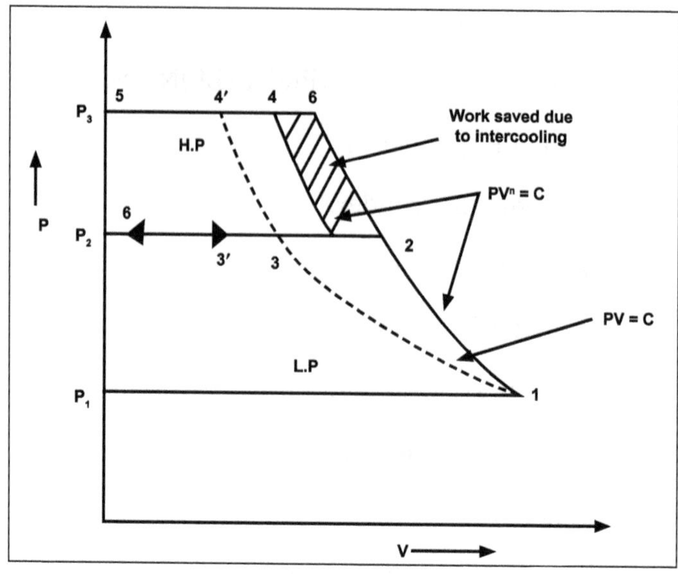

Isothermal curve.

Minimum work required for a two stage compressors with inter cooling: When the inter cooling is perfect, the maximum saving in work would be obtained,

$$W = \frac{n}{n-1}(P_1V_1)\left[\left(\frac{P_2}{P_1}\right)^{\frac{n-1}{n}} + \left(\frac{P_4}{P_3}\right)^{\frac{n-1}{n}} - 2\right] \quad ...(i)$$

$$\text{Put } \frac{n}{n-1} = k$$

$$W = \text{Constant x}\left[\left(\frac{P_2}{P_1}\right)^k + \left(\frac{P_3}{P_2}\right)^k - 2\right]$$

For minimum work,

$$\frac{dW}{dP_2} = 0$$

$$\frac{dW}{dP_2} = \text{Constant}\left[\frac{KP_2^{k-1}}{P_1^k} - \frac{KP_3^k}{P_2^{k+1}}\right] = 0$$

$$\frac{P_2^{\,k-1}}{P_1^{\,k}} = \frac{P_3^{\,k}}{P_2^{\,k+1}}$$

$$P_2^{\,(k-1)}P_2^{\,(k+1)} = P_1^{\,k}P_3^{\,k}$$

$$(P_2)^{2k} = (P_1.P_3)^k$$

$$P_2^2 = P_1.P_3 \text{ (Or) } \frac{P_2}{P_1} = \frac{P_3}{P_2} \text{ (Or) } P_2 = \sqrt{P_1P_3} \text{ (2)}$$

Substitute (2) in (1),

$$W = \frac{n}{n-1}(P_1V_1)\left[\left(\frac{P_2}{P_1}\right)^{\frac{n-1}{n}} + \left(\frac{P_2}{P_1}\right)^{\frac{n-1}{n}} - 2\right]$$

$$W = \frac{2n}{n-1}(P_1V_1)\left[\left(\frac{P_2}{P_1}\right)^{\frac{n-1}{n}} - 1\right] \text{ Joules(3)}$$

$$\frac{P_2}{P_1} = \frac{P_3}{P_2}$$

$$\left(\frac{P_2}{P_1}\right)^2 = \frac{P_3}{P_2} \text{ x } \frac{P_2}{P_1} = \frac{P_3}{P_1}$$

$$\frac{P_2}{P_1} = \left(\frac{P_3}{P_1}\right)^{\frac{1}{2}}$$

Then equation (iii),

$$W = \frac{2n}{n-1}P_1V_1\left[\left(\frac{P_2}{P_1}\right)^{\frac{n-1}{n}} - 1\right]$$

Putting $P_1V_1 = mRT_1$ in equation,

$$W = \frac{2n}{n-1}mRT_1\left[\left(\frac{P_3}{P_1}\right)^{\frac{n-1}{2n}} - 1\right]$$

For x stage,

$$W = \frac{xn}{n-1}mRT_1\left[\left(\frac{P_3}{P_1}\right)^{\frac{n-1}{2n}} - 1\right]$$

Inter Cooler

If an inter cooler is installed between cylinders, in which the compressed air is cooled between cylinders then the final delivery temperature is reduced. This reduction in temperature means a reduction in internal energy of the delivered air and since this energy must have from the input energy required to drive the machine, this results in a decrease in input requirement for a given mass of delivered air.

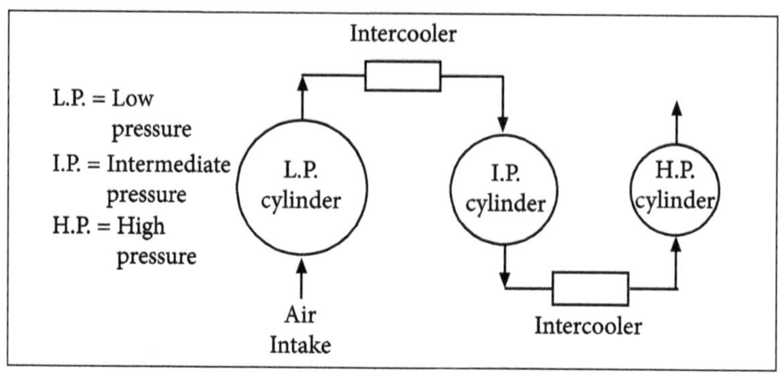

Three - stage compressor.

Effects of inter-cooling in a multi-stage compression process

- The work done per day of air is reduced in multistage compression with inter cooler as compared to single stage compression for the same delivery pressure.
- It improves the volumetric efficiency for the given pressure ratio.
- It provides effective lubrication because of lower temperature range.
- It reduces the cost of compressor.

Need and Advantages of Multistage Compression with a Pv Diagram

Need

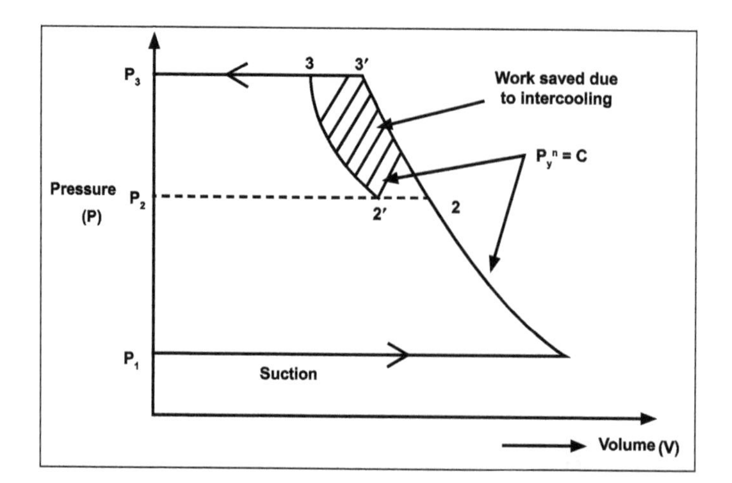

In multistage compressor, the compression of the air from the initial pressure to the final pressure is carried out in more than one cylinder. It is used to get high pressure air. In a compressor, when compression ratio exceeds five, generally multistage compression is adopted.

Advantages of multistage air compressor:

- Work done per kg of air is reduced in multistage compression with inter cooler as compared to single stage compression for the same delivery pressure.

- Better mechanical balance can be achieved with multistage compressors.
- It reduces the leakage loss considerably.
- Volumetric efficiency is improved by increasing number of Stages.
- It gives more uniform torque and hence a smaller size flywheel is required.
- Lower operating temperature permits the use of cheaper materials for construction.

Condition for Minimum Work Input In a Multistage Compression Process

Derive the necessary condition for minimum work input in a multistage compression process.

Work done by two-stage reciprocating air compression is given by,

$$W = \frac{n}{n-1} P_1 V_1 \left[\left(\frac{P_2}{P_1} \right)^{\frac{n-1}{n}} + \left(\frac{P_3}{P_2} \right)^{\frac{n-1}{n}} - 2 \right]$$

Differentiating with respect to P_2 and equating to zero,

$$\frac{dW}{dP_2} = 0$$

$$\frac{d}{dP_2} \left[\frac{n}{n-1} P_1 V_1 \left[\left(\frac{P_2}{P_1} \right)^{n-1/n} + \left(\frac{P_3}{p_2} \right)^{n-1/n} - 2 \right] \right] = 0$$

Let, $\frac{n}{n-1} = K$ (constant),

$$\frac{d}{dP_2} \left[kP_1 V_1 \left[\left(\frac{P_2}{P_1} \right)^k + \left(\frac{P_3}{P_2} \right)^k - 2 \right] \right] = 0$$

$$kP_1 V_1 \frac{d}{dp_2} \left[\left(\frac{P_2}{p_1} \right)^k + \left(\frac{P_3}{P_2} \right)^k - 2 \right] = 0$$

$$kP_1 V_1 \left[\left(\frac{1}{P_1} \right)^k k(P_2)^{k-1} + (P_3)^k (-k)(P_2)^{-k-1} \right] = 0$$

$$\left[\left(\frac{1}{P_1} \right)^k k(P_2)^{k-1} + (P_3)^k (-k)(P_2)^{-k-1} \right] = 0$$

$$k(P_1)^{-k}(P_2)^{k-1} - k(P_3)^{k}(P_2)^{-k-1} = 0$$

$$k(P_1)^{-k}(P_2)^{k-1} = k(P_3)^{k}(P_2)^{-k-1}$$

$$\frac{(P_2)^{k-1}}{(P_2)^{-k-1}} = \frac{(P_3)^{k}}{(P_1)-k}$$

$$(P_2)^{k-1} \times (P_2)^{k+1} = (P_3)^{k} \times (P_1)^{k}$$

$$(P_2)^{2k} = (P_3P_1)^{k} \text{ (or)}$$

$$P_2^2 = P_1P_3$$

$$\therefore P_2 = \sqrt{P_1P_3}$$

2.7.1 Optimum Intercooler Pressure, Performance and Design Calculations of Reciprocating Compressors

$$W = W_1 + W_2$$

Where,

W_1 is the work done in the low pressure stage and W_2 is the work done in the high pressure stage.

$$W = \frac{mRn(T_2 - T_1)}{(n-1)} + \frac{mRn(T_6 - T_5)}{(n-1)}$$

Since,

$$T_2 = T_1\left(\frac{p_2}{p_1}\right)^{1-1/n} \text{ and } T_6 = T_5\left(\frac{p_6}{p_5}\right)^{1-1/n}$$

Then assuming the same value of n for each stage,

$$W = mR\left[\left\{\frac{nT_1}{(n-1)}\right\}\left\{\frac{p_2}{p_1}\right\}^{1-(1/n)} - 1\right] + mR\left[\left\{\frac{nT_6}{(n-1)}\right\}\left\{\frac{p_6}{p_5}\right\}^{1-(1/n)} - 1\right]$$

Since,

$$P_2 = P_5 = P_M \text{ and } P_6 = P_H \text{ and } P_1 = P_L$$

$$W = mR\left[\left\{\frac{nT_1}{(n-1)}\right\}\left\{\frac{p_M}{p_L}\right\}^{1-(1/n)} - 1\right] + mR\left[\left\{\frac{nT_6}{(n-1)}\right\}\left\{\frac{p_H}{p_M}\right\}^{1-(1/n)} - 1\right]$$

For minimum value of W, we differentiate with respect to P_M and equate to zero.

$$\frac{dW}{dp_M} = mRT_1 p_L^{(1-n)/n} p_M^{-1/n} - mRT_5 p_H^{(1-n)/n} p_M^{-1/n}$$

If the intercooler returns the air to the original inlet temperature so that $T_1 = T_5$, then equating to zero reveals that for minimum work,

$$p_M = (P_L P_H)^{1/2}$$

It can further be shown that when this is the case, the work done by both stages is equal. When K stages are used, the same process reveals that the minimum work is done when the pressure ratio for each stage is $(p_L/p_H)^{1/K}$.

2.7.3 Air Motors

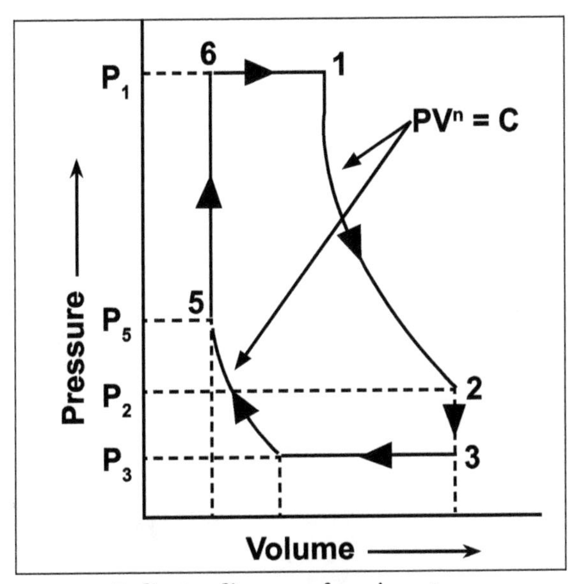

Indicator diagram of an air motor.

Reciprocating air motors are used where the use of electric motor or I.C. engine is not permissible, as in mining engineering. It works on the cycle which is the reverse of the reciprocating compressor cycle. The indicator diagram for a reciprocating air motor is shown in the figure below.

The high pressure air from the supply main is applied to the cylinder of air motor through the valves at pressure P_1. The air expands poly-tropically from 1 to 2. The exhaust valve opens and the air is released at constant volume from 2 to 3 to pressure P_b which is the back pressure.

Then the air is exhausted along the line 3-4 at constant pressure. At point 4 the exhaust valve closes and the polytrophic compression of trapped air takes place along 4 to 5.

At point 5 the admission valve opens and the pressure in the cylinder rises rapidly to the inlet pressure P_1. Again air is supplied at constant pressure up to the point of cutoff 1.

The work done is given by,

$$W=P_1(V_1-V_6)+\frac{P_1V_1-P_2V_2}{n-1}-P_3(V_3-V_4)-\frac{P_5V_5-P_4-V_4}{n-1}J$$

Problems

1. A 2 kg/s of air enter the LP cylinder of a two stage, reciprocating air compressor. The overall pressure ratio is 9. The air at inlet to compressor is at 100 Kpa and 35°C. The index of compression in each cylinder is 1.3. Let us determine the intercooler pressure for perfect inter-cooling. Let us also find the minimum power required for compression and percentage saving over single stage compression. Take R = 0.287 kJ/kg°K and Cp = 1 kJ/kg°K.

Solution:

Given:

$$m = 2\ kg/S$$

$$\frac{P_3}{P_1}=9$$

$$P_1 = 100Kpa$$

$$P_3 = 900Kpa$$

$$T_1 = 35°C + 273 = 308\ k$$

$$n = 1.3$$

$$P_2 = \sqrt{P_1P_3} = \sqrt{100 \text{ x } 900}$$

$$= 300\ kPa$$

Work done for x number of stages,

$$W=\frac{xn}{n-1}P_1V_1\left[\left(\frac{P_{n+1}}{P_1}\right)^{\frac{n-1}{nx}}-1\right]$$

$$x = 2$$

$$=\frac{2n}{n-1}mRT_1\left[\left(\frac{P_3}{P_1}\right)^{\frac{n-1}{2n}}-1\right]$$

$$=\frac{2 \text{ x } 1.3}{1.3-1} \text{ x } 2 \text{ x } 0.287 \text{ x } 308\left[(9)^{\frac{1.3-1}{2 \text{ x } 1.3}}-1\right]$$

$P_1 = 442.13$ kW

Work done for single stage,

$$W=\frac{n}{n-1}P_1V_1\left[\left(\frac{P_2}{P_1}\right)^{\frac{n-1}{n}}-1\right]$$

$$=\frac{n}{n-1}mRT_1\left[\left(\frac{P_2}{P_1}\right)^{\frac{n-1}{n}}-1\right]$$

$P = P_2$ In a single stage,

$$W=\frac{1.3}{1.3-1} \text{ x } 2 \text{ x } 0.287 \text{ x } 308\left[(9)^{\frac{1.3-1}{1.3}}-1\right]$$

$P_2 = 505.9$ kW

Saving in Power = 505.9 - 442.73

= 63.77 kW

$$\% \text{ of saving power } =\frac{P_2-P_1}{P_2}$$

$$=\frac{63.77}{505.9}=12.6\%$$

2. In an open cycle gas turbine plant air enters the compressor at 1 bar and 27°C. The pressure after compression is 4 bar. The isentropic η_s of the turbine and compressor are 85% and 80% respectively. Air fuel ratio is 80:1. CV of the fuel used is 42000 kJ/kg. Mass flow rate of air is 2.5 kg/s. Let us determine the power output from the plant and the cycle η . Assume C_P and γ to be same for both air and products of combustion.

Solution:

Given:

P_1= 1bar , T_1 = 300K , P_2= 4bar , η_c =80%

η_T=85%

A/F = 80:1 C_V= 42000kJ/kg m_a= 2.5kg/s

$C_p = 1.005$

$\gamma = 1.4$

To find:

- Power output.
- η_{cycle}.

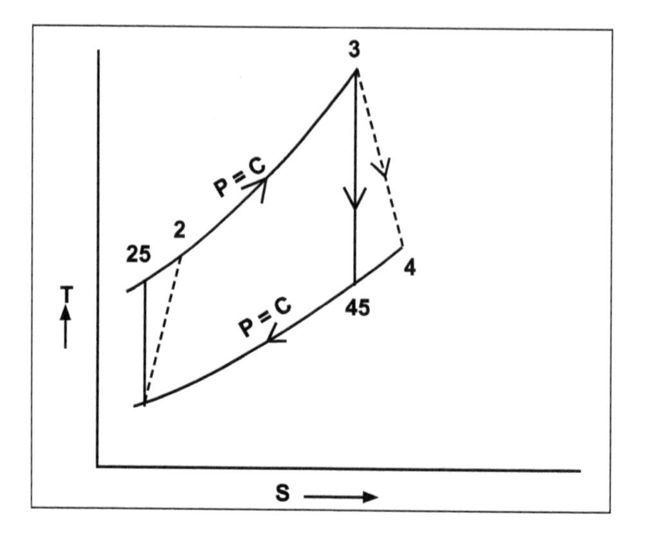

Process 1-2s is Isentropic

$$\text{i.e., } \frac{T_{2s}}{T_1} = \left(\frac{P_2}{P_1}\right)^{\frac{\gamma-1}{\gamma}} \quad \therefore T_{2s} = 300(4)^{0.286} = 445.8\text{K}$$

But,

$$\eta = \frac{T_{2s} - T_1}{T_2 - T_1} \quad \text{i.e., } 0.8 = \frac{445.8 - 300}{T_2 - 300} \quad \therefore T_2 = 482.3\text{K}$$

Heat Addition Rate During Process 2-3

$$\dot{Q}_H = \dot{m}_f CV = \left(\dot{m}_a + \dot{m}_f\right) C_P \left(T_3 - T_2\right)$$

$$\text{But, } \dot{m}_f = \dot{m}_a \frac{\dot{m}_f}{\dot{m}_a} = \frac{2.5}{80} = 0.03125\,\text{kg/s}$$

$$\therefore \dot{Q}_H = 0.03125(42000)$$

$$= 1312.5$$

$$= (2.5 + 0.03125)1.005(T_3 - 482.3)$$

Process 3-4s is Isentropic

$$\text{i.e., } \frac{T_{4s}}{T_3} = \left(\frac{P_4}{P_3}\right)^{\frac{\gamma-1}{\gamma}}$$

$$\therefore T_{4s} = 998.2\left(\frac{1}{4}\right)^{0.236} = 671.6\text{K}$$

$$\text{But } \eta_T = \frac{T_3 - T_4}{T_3 - T_{4S}} \quad \text{i.e., } \quad 0.85 = \frac{998.2 - T_4}{998.2 - 671.6} \quad \therefore T_4 = 720.6\text{K}$$

$$\eta_{cycle} = \frac{W_{net}}{Q_H} = \frac{247.6}{1312.5} = 18.9\%$$

3

Steam: Generator and Nozzles

3.1 Properties of Steam

1. Enthalpy or Total Heat of Water, h_f

It is defined as the quantity of heat required to raise the temperature of one kg of water from 0°C to its boiling point or saturation temperature corresponding to the pressure applied. It is also called enthalpy of saturated water or liquid heat and is represented by h_f.

$$h_f = \text{Specific heat of water, } c_{pw} \times \text{Rise in temperature}$$
$$= 3.187 \text{ k } \Delta t, \text{ kJ / kg}$$

2. Latent Heat of Steam, h_{fg}

Latent heat of steam at a particular pressure may be defined as the quantity of He required converting one kg of water at its boiling point into dry saturated steam at the same pressure. It is denoted by h_{fg}.

3. Dryness Fraction

It is defined as the ratio of mass of dry steam actually present to the mass of wet steam. It is denoted by x.

Let ,

m_s = Mass of dry steam, kg

m_w = mass of water vapour in steam, kg

x = Dryness fraction of the sample,

Then $x = \dfrac{m_s}{m_s + m_w}$

For dry steam,

$$m_w = 0 \text{ and } x = 1$$

The quality of steam is the dryness fraction expressed as a percentage.

$$\text{Quality of steam} = 100x$$
$$\text{Wetness fraction} = 1 - x = m_w / (m_s + m_w)$$

The wetness fraction expressed in percentage is called priming.

$$\text{Priming} = 100\ (1 - x)$$

4. Enthalpy of Wet Steam

It may be defined as the quantity of heat required to convert one kg of water at 0°C, at constant pressure, into wet scheme.

$$h = h_f + xh_{fg}$$

When steam is dry saturated, then x=1 and,

$$h_g = h_f + h_g$$

5. Total Enthalpy of Superheated Steam

Let,

c_{PS} = Specific heat of superheated steam

t_s = Temperature of formation of steam, °C

t_{sup} = Temperature of formation of steam, °C

Then heat of superheat,

$$= c_{ps}\left(t_{sup} - 1\right)$$

$$h_{sup} = \left(h_f + h_{fg}\right)c_{ps}\left(t_{sup} - t_s\right)$$

$$= h_g + c_{ps}\left(t_{sup} - t_s\right)$$

3.1.1 Measurement of Dryness Fraction

I) Measurement of Dryness Fraction using Separating Calorimeter

In this case, the dryness fraction of the steam is measured by separating the water particles from the wet steam.

The wet steam from the boiler enters the top of the calorimeter through a control valve. Then it strikes a perforated cup and in turn it undergoes direction reversal. Due to higher moment of inertia, water particles in the wet steam get separated. The separated water collects at the bottom of the inner chamber and its amount can be measured by using the water gauge.

It is denoted by,

Dryness fraction, x = Mass of dry stream / Mass of wet stream

$$x = \frac{m_s}{m_s + m_w}$$

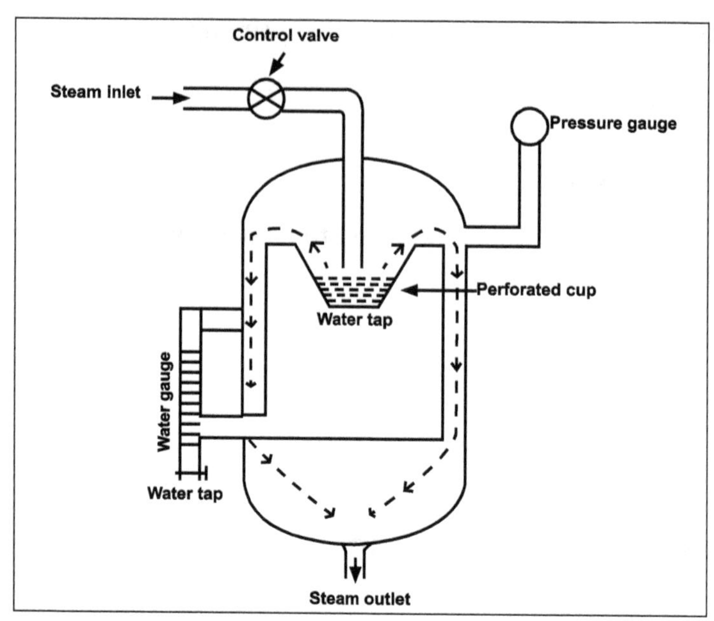

Separating Calorimeter.

Where,

m_g – mass of dry steam in kg.
m_f – mass of water vapour in suspension.
This term is applicable only for wet steam.

For dry steam m o

ii) Measuring of Dryness Fraction Using Throttling Calorimeter

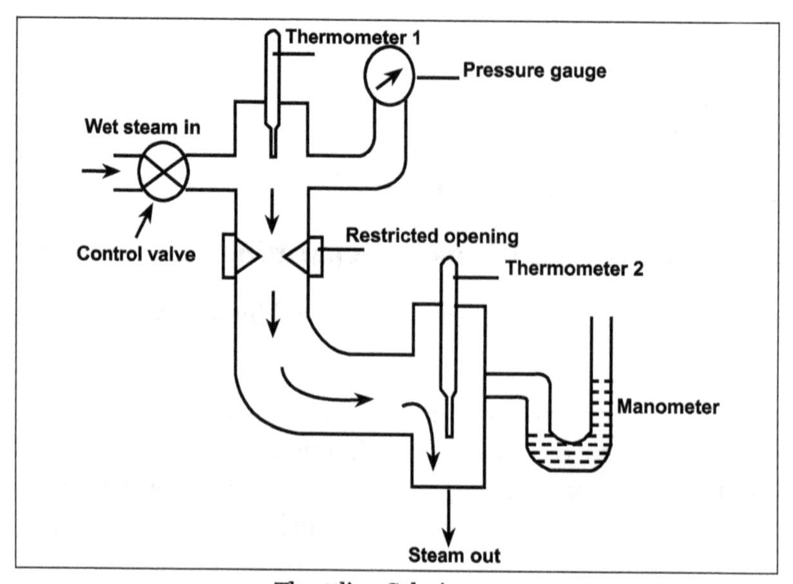

Throttling Calorimeter.

The wet steam is throttled by passing it through a narrow opening i n its passage. The wet steam from the steam generator, i.e. the boiler, is admitted through a control valve to a restricted opening. The temperature and pressure of the wet steam is measured at this section by using a thermometer and a pressure gauge. When the wet steam passes through a restricted opening, the total heat of steam remains at a constant level.

Due to throttling action, the wet vapour condenses into liquid water and the pressure of the wet steam reduces. The temperature of the steam leaving the chamber is measured by means of a thermometer. The pressure of steam after throttles can be measured by using a manometer. After throttling, the dry steam is available in superheated region.

Let, p_1 = Pressure of stream before throttling.

T_1 = Temperature of steam before throttling.

h_{f1} = Sensible heat of water i.e. fluid enthalpy at state 1.

h_{fg1} = Latent heat at state 1.

h_{g2} = Total heat of dry steam at state 2 i.e. gas enthalpy at state 2.

t_{sat} = saturation temperature corresponds to pressure p_2,

t_2 = Temperature of steam at state 2.

Using the energy balance equation for a throttling process,

$h_1 = h_2$ i.e. total enthalpy before throttling is equal to total enthalpy after throttling.

$$h_{f1} + x.\,h_{fg1} = h_{g2} + C_p(t_2 - t_{sat}),$$

Using this equation the dryness fraction of entering wet steam can be calculated.

3.2 Use of Steam Table

The different properties of steam, i.e. saturation temperature, specific volume, sensible heat, latent heat, total heat and entropy corresponding to various pressures can be obtained by performing experiments. The value of these properties is gathered in a tabular form and is known as steam tables. These properties may be tabulated either on pressure basis or on temperature basis and correspond to 1 kg of dry and saturated steam.

1. Sensible Heat (h)

The quantity of heat required in kJ to raise the temperature of 1 kg of water from 0°C to the boiling point temperature corresponding to the given pressure, is known as sensible heat of water. It is denoted by the letter h and is also called total heat (or enthalpy) of water.

2. Latent Heat of Vaporization (h_{fg})

The quantity of heat required in kJ to convert 1 kg of water at its saturation temperature for a given pressure. To 1 kg of dry saturated steam is known as latent heat of vaporization. It is denoted by the letter h_{fg}.

3. Enthalpy (or) Total Heat of Wet Steam

The quantity of heat required to convert 1 kg of water at 0°C into wet steam at constant pressure is known as total heat (or enthalpy) of wet steam. It is denoted by,

$$h = h_f + x \cdot h_{fg}$$

Where,

h_f - Fluid enthalpy

h_{fg} - Latent heat = $h_g - h_f$.

4. Enthalpy (or) Total Heat of Dry Saturated Steam

The quantity of heat required to convert 1kg of water at 0°C into dry saturated steam at constant pressure is known as total heat of dry saturated steam. It is denoted by the letter (h)

$$h = h_g$$

Where,

h_g -Enthalpy or total heat at the saturation point at that pressure.

5. Total Heat of Superheated Steam

The quantity of heat required to convert I kg of water at 0°C at constant pressure into superheated steam is known as total heat of superheated steam. It is denoted by,

$$h = h_g + C_p\left(T_{sup} - T_{sat}\right)$$

Where,

h_g -Enthalpy of saturated vapour at that pressure.

C_p- Specific heat at constant pressure for the working substance.

T_{sup} -Superheated temperature and T_{sat} is the saturated temperature and (T_{sup} -T_{sup}) is known as degree of superheat.

The average value of C_p for saturated steam is 2.0934 kJ/kg . K.

6. Specific Volume of Dry Steam

The volume of I kg of dry steam at a given pressure is known specific volume of dry steam. It is denoted by the letter v. Its values corresponding to different pressures are given in steam tables. The value of v decreases with increase in pressure. Mathematically,

$$v = v_g m^3 / kg$$

Density of dry steam = 1/v — kg/m³ .

7. Specific Volume of Wet Steam

The volume of one kg of wet steam is equal to the volume of dry portion of steam plus the volume of water in suspension. It is denoted by the letter v.

Let, x be the dryness fraction of wet steam.

Let 1kg of wet steam will consist of m_v kg of dry steam and m_f kg of water in suspension.

Specific volume of wet steam = Volume of dry steam + Volume of water in suspension Where,

$$v = v_f + x.v_{fg}$$

$$v_{fg} = v_g - v_f$$

As volume of water at low pressure is very small the term v_f can be neglected.

Therefore, specific volume of wet stream = $x.v_g$ m^3/kg .

8. Specific Volume of Superheated Stream

The volume of superheated steam may be found out on the assumption that superheated steam behaves as a perfect gas as soon as the process of superheating begins. i.e. from dry saturated condition.

By applying gas law to steam at the beginning of superheating and at the end of superheating we have, since

$$\frac{p_{sat} \times v_{sat}}{T_{sat}} = \frac{p_{sup} \times v_{sup}}{T_{sup}}$$

$$p_{sat} = p_{sup}$$

$$\frac{v_{sat}}{T_{sat}} = \frac{v_{sup}}{T_{sup}}$$

$$\therefore \quad v_{sup} = \frac{T_{sup}}{T_{sat}} \times v_{sat}$$

9. Entropy

The entropy of water at 0°C is taken as zero. The water is heated and evaporated at constant pressure. The steam is also superheated at constant pressure in super heaters. For this reason the entropy of steam can be calculated from the formula for the change of entropy at constant pressure. The entropy of wet steam, dry steam and superheated steam is obtained as given below.

10. Entropy of Wet Steam

Consider 1kg of wet steam with dryness fraction x. The entropy of wet steam reckoned above the freezing point of water (0°C) is given by,

$$s = s_f + x \cdot s_{fg}$$

Where,

$$s_{fg} = s_g - s_f \, kJ/kg \cdot K$$

11. Entropy of Dry Steam

Consider 1 kg of dry saturated steam. The entropy of dry saturated steam reckoned above the freezing point of water (0°C) is given by,

$$s = s_g$$

12. Entropy of Superheated Steam

Consider 1 kg of dry saturated steam. The steam is superheated at constant pressure.

$$s = s_g + C_P \ln\left(\frac{T_{sup}}{T_{sat}}\right)$$

For steam, the value of C_p varies between 2 kJ/kg . K and 2.3 kJ/kg . K.

In general the value of C_p for steam can be taken as 2.0934 kJ/kg. K.

3.2.1 Mollier Chart-T-S and H-S Diagrams for Representing

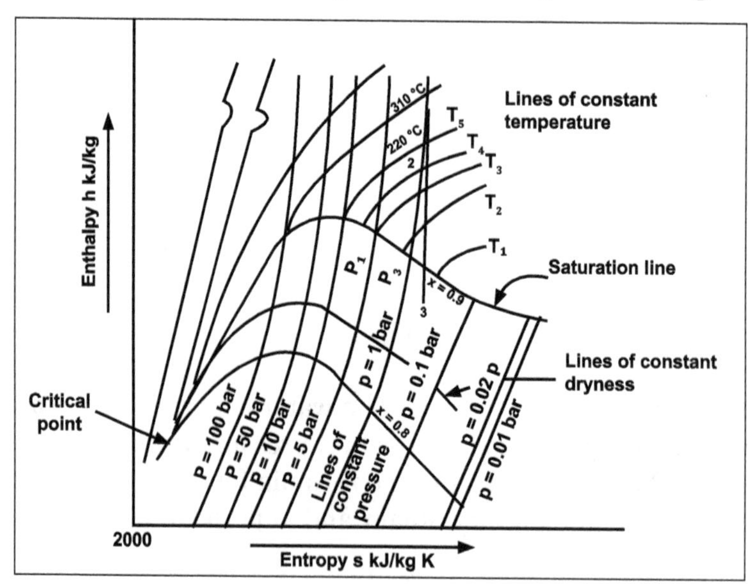

Enthalpy- entropy (h-s) diagram.

The Mollier diagram is a very commonly used chart by engineers, in which vertical ordinate represents the enthalpy while the base represents the entropy. A Mollier diagram for steam is flown in the figure. in which the regions above and below the saturation line represent the superheated and wet conditions of steam respectively. The lines of constant dryness fraction are shown in wet steam region while the lines of constant temperature arc in the region of superheat.

In order to facilitate taking readings horizontal as well as vertical line arcs drawn at small intervals. An adiabatic process will be represented by a vertical line on the Mollier diagram while expansion during which the total heat remains constant will be represented by a horizontal line. One of the outstanding advantages of this diagram is that the drop in the total heat during an adiabatic expansion can be directly read from it.

- Lines of constant pressure are indicated by p_1, p_2, etc. and lines of constant temperature by T_1, T_2, etc.
- Any two independent properties which appear on the chart are sufficient to define the state for example, p_1. x_1 define state 1 and h_1 can be read off the vertical axis.
- In the superheat region, pressure and temperature can define the state. For example, p_3 and T_4 define the state 2 and h_2 can be read off.
- A line of constant entropy between two state points 2 and 3 define the properties at all points during an isentropic process between the two states.

h-s Diagram

These entropy changes are represented in the figure (a). The curve 1, 2, 3, 4, 5, 6 is the isobar of I bar. Repeat the above procedure to construct an isobar of 2 bar, a similar kind of curve will be obtained.

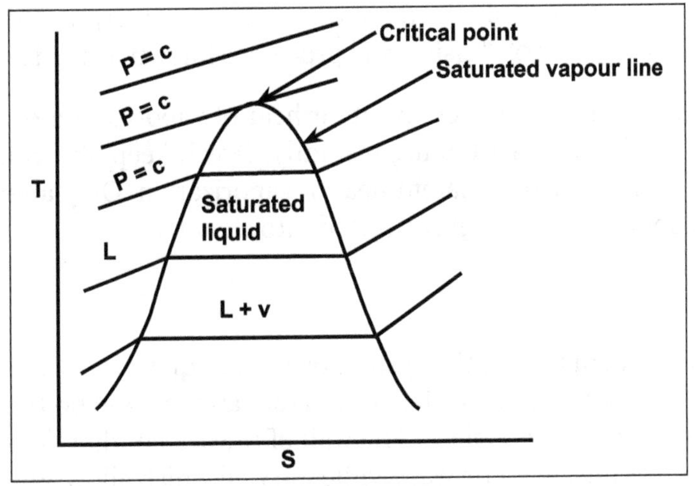

Entropy changes.

If the corresponding points on different isobars are connected by means of a smooth curve, the phase equilibrium diagram on T-s plot will be obtained. Most often the liquid-vapour transformations only are of interest, hence in the figure (b), shows the liquid, vapour and transition zones in T-s plot.

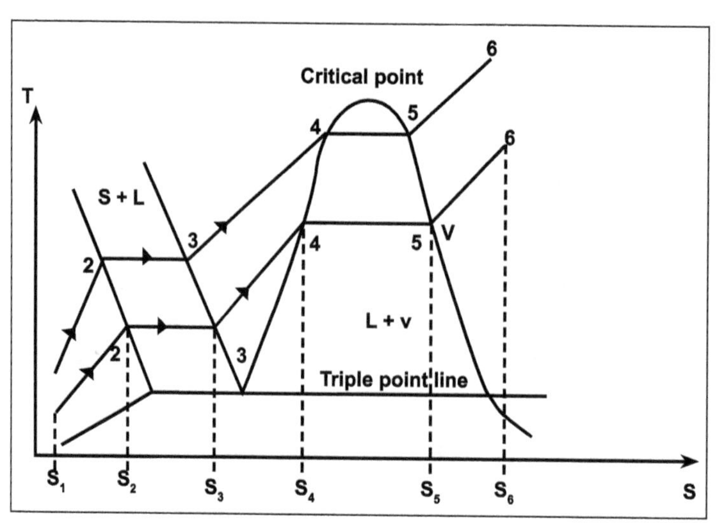

Liquid, vapour and transition zones.

3.2.2 Thermodynamic Processes

When system undergoes the change from one thermodynamic state to final state due change in properties such as temperature, pressure, volume etc., the system is considered to have undergone thermodynamic process. Several types of thermodynamic processes are as follows. Such as isothermal process, adiabatic process, isochoric process, isobaric process, and reversible process. These have been described below:

1. Isothermal Process

When the system undergoes the change from one state to the other, but if its temperature remains constant, the system is said to have undergone an isothermal process. For instance, consider our example of hot water in thermos flask, if we remove certain quantity of water from the flask, but keep its temperature constant at 50° Celsius, the process is known as the isothermal process.

Another example of an isothermal process is latent heat of vaporization of water. When we heat water to 100°Celsius, it will not start boiling instantly. It will keep absorbing heat at a constant temperature and this heat is called the latent heat of vaporization. Only after absorbing this heat, water at constant temperature, it will get converted into steam.

2. Adiabatic Process

The process, in which heat content of the system or certain quantity of the matter remains constant, is known as the adiabatic process. Hence, an adiabatic process no transfer of heat between the system and its surroundings take place. The wall of the system that does not allow the flow of heat through it is known as the adiabatic wall, while the wall which allows the flow of heat is known as the diathermic wall.

3. Isochoric Process

The process, in which the volume of the system remains constant, is known as the isochoric process. Heating of gas in a closed cylinder is an example of the isochoric process.

4. Isobaric Process

The process in which the pressure of the system remains constant is known as the isobaric process. An example: Suppose there is fuel in piston and cylinder arrangement. When this fuel is burnt the pressure of the gases is generated inside the engine and since more fuel burns more pressure is created. But when the gases are allowed to expand by allowing the piston to move outside, the pressure of the system may be kept constant. The constant pressure and volume processes are really important. The Otto and diesel cycle are used in the petrol and the diesel engine respectively. They have constant volume and constant pressure processes. In practical situations ideal constant pressure and constant pressure processes may not be achieved.

5. Reversible Process

In simple words the process which can be reversed back completely is known as a reversible process. This means the final properties of the system should be perfectly reversed back to original properties. The process may be perfectly reversible only if the changes in the process are infinitesimally small. In practical situations it is impossible to trace these extremely small changes in extremely small time, thus the reversible process is also an ideal process. The changes that occur during the reversible process are in equilibrium with each other.

1. Let us consider the Steam flows through a small turbine at the rate of 5000 kg/hr entering at 15 bar, 300°C and leaving at 0.1 bar with 4% moisture. The steam enters at 80 m/sec at a point 2 m above the discharge and leaves at 40 m/sec compute the shaft power assuming that the device is adiabatic but considering kinetic and potential energy changes. Calculate the diameters of the inlet and discharge tube.

Solution:

Given:

p_1 = 15 bar, 300°C

p_2 = 0.1 bar, x = 0.96

C_1 = 80 m/s, C_2 = 40 m/sec.

Z_1 = 2 m

m = 5000 Kg/hr = 1.389 Kg/sec

h_1 at 15 bar and 300°C = 3038.9 kJ/Kg

= 3038.9×10^3 J/Kg

h_2 at 0.1 bar and x = 0.96

$$gZ1+\frac{C_1^2}{2}+h_1 = gZ_2^0+\frac{C_2^Z}{2}+h_2+w$$

$$2 \text{ x } 9.81+\frac{80^2}{2}+3038.9 \text{ x } 10^3 = 0+\frac{40^2}{2}+2488.984 \text{ x } 10^3+W$$

$W = 552.335\ KJ/Kg$

Power required $= 552.335 \times 1.3889$

$= 767.13\ kW$

Mass Flow rate

$$\dot{m} = \frac{a_1 C_1}{v_1}$$

$v_1 = v_{sup}$ at 15 bar and 300°C

$= 0.1697\ m^3/Kg$

$$1.3889 = \frac{a_1 \times 80}{0.1697}$$

$a_1 = 2.9462 \times 10^{-3} m^2$

Inter diameter = 61.2 mm

$$\dot{m} = \frac{a_2 C_2}{v_2} \quad \dot{v}_2 = 14.088\, m^3/kg$$

$$1.3889 = \frac{a_2 \times 40}{14.088}; \quad a_2 = 0.489\, m^3$$

Discharge dia = 789 mm.

2. If water is at 65°C at 1 atm. Let us find the state of water and its specific enthalpy.

Solution:

Corresponding to 65°C, saturation pressure

$P_{Sat} = 25.03$ hpa

Since $Pc_{ut} < P$, the water is steam which is in super-heated state.

Specific enthalpy, h = 2621.07 kJ / kg

$h_2 = h_f + x\, h_{fg}$

$= 191.8 + 0.96\,(2392.9)$

$= 2488.984$ KJ/Kg (or) 2488.9×10^3J/sec

3. Let us consider the Steam expands isentropically in a nozzle from IMPa, 250°C to 10 Kpa. The steam flow rate is 1 Kg/sec. Let us find the velocity of steam at the exit from the nozzle and the exit area of the nozzle, neglecting the velocity of steam at inlet to the nozzle. The exhaust steam from the nozzle flows into a condenser and flows out as saturated water. The cooling water enters the condenser at 25°C and leaves at 35°C. Determine the mass flow rate:

Solution:

Given:

$$p_1 = 1\text{ Mpa} = 1000\text{ Kpa} = 10\text{ bar};\ T_1 = 250°C$$
$$p_2 = 10\text{ Kpa} = 0.1\text{ bar}$$
$$m = 1\text{ Kg/sec}$$
$$h_1 = h_{sup}\text{ at 10 bar and } 250°C = 294 \times 10^3\text{ J/Kg.}$$

$$S1 = S2$$

$$6.926 = S_f + xS_{fg}$$
$$= 0.649 + x(7.502)$$
$$x = 0.836$$

$$h_2 = h_f + x\,h_{fg}$$

$$= 199.7 + 0.836\,(2388.4)$$

$$= 2196.4\text{ kJ/Kg}$$

$$gZ1 + \frac{C_1^2}{2} + h_1 = gZ_2 + \frac{C_2^2}{2} + h_2$$

$$C_2 = \sqrt{2(h_1 - h_2) + S_1^{270}}$$

$$C_2 = \sqrt{2[2943 \text{ x } 10^3 - 2196.4 \text{ x } 10^3 1]}$$

$$C_2 = 1221.90\text{ m/sec}$$

$$\dot{m} = \frac{a_2 C_2}{V_2}$$

$$1 = \frac{a_2 \text{ x } 1221.96}{12.26}$$

$$a_2 = 0.01\text{ m}^2$$

In condenser

$$\dot{m}\left[h_2 - h_f \text{ at } 0.1 \text{ bar}\right] = \left[m_W C_{pW}\left[T_2 - T_1\right]\right.$$

$$1\left[2196.4 - 191.8\right] = m_W \times 4.1867\left[35 - 25\right]$$

$$m_w = 47.87 \text{ Kg/sec}$$

4. Let us consider a large insulated vessel is divided into two chambers, one centering 5 kg of dry saturated steam at 0.2 MPa and the other 10 kg of steam, 0.8 quality at 0.5 MB. If the partition between the chambers is removed and the steam is mixed thoroughly and allowed to settle, Find the final pressure, steam quality and entropy change in the process.

Solution:

Given:

$$\text{At } \Delta S = 0.43 \text{ kJ/kg} 0.2 \text{ MPa}, U_g = U_i = 0.8857 \text{ m}^3/\text{kg}$$

$$V_1 = M_1 U_1 = 4.4285 \text{ m}^3$$

$$\text{At } 0.5 \text{ MPa}, U_2 = U_f + x_2 U_{fg} = 0.30101 \text{ m}^3/\text{kg}$$

$$V_2 = m_2 U_2 = 3.0101/\text{m}^3$$

$$V_m = V_1 + V_2 = 7.4386 \text{ m}^3$$

At 0.2 MPa,

$$h_g = h_1 = 2706.7 \text{ kJ/kg}$$

$$x_1 = h1 - P_1 U_1 = 2706.7 \text{ kj/kg}$$

$$h_2 = 2323.03 \text{ kJ/kg} ; h_3 = 2453.6 \text{ kJ/kg}$$

$$V_3 = 0.496 \text{ m}^3/\text{kg} ; x_3 = 0.87 ; S_3 = 6.29 \text{ kJ/kg k}$$

$$S_2 = S_{f2} + 0.8\, S_{fg2} = 5.8292 \text{ kJ/kg K}$$

$$\Delta S = m_3 S_3 - \left(m_1 S_1 + m_2 S_2\right)$$

$$\Delta S = 0.43 \text{ kJ/kg}$$

5. Let us consider a reheat cycle, the initial steam pressure and the maximum temperature with 150 bar and 550°C respectively. If the condenser pressure is 0.1 bar and the moisture at the condenser

inlet is 5% and assuming ideal processes, Determine (1) reheat pressure, (2) cycle efficiency and (3) the steam rate.

Solution:

Given:

$$P_1 = 150 \text{ bar} \; ; \; T_1 = 823 \text{ k} \; ; \; P_3 = 0.1 \text{ bar} \; ; \; x_3 = 0.95$$

From steam table at 0.1 bar,

$h_{f3} = 191.8$ kJ/kg ; $h_{fg3} = 2392.9$ kJ/kg

$S_{f3} = 0.649$ kJ/kg k

$S_{fg3} = 7.602$ kJ/kg k; $U_{f3} = 0.001010$ m³/kg

$S_3 = S_{f3} + x_3 \times S_{fg3} = 0.649 + 0.95 \times 7.502 = 7.7759$ kJ/kgk

$h_3 = h_{f3} + x_3 x h_{fg3} = 191.8 + 0.95 \times 2392.9 = 2465.06$ kJ/kg

Using Mollier's Chart:

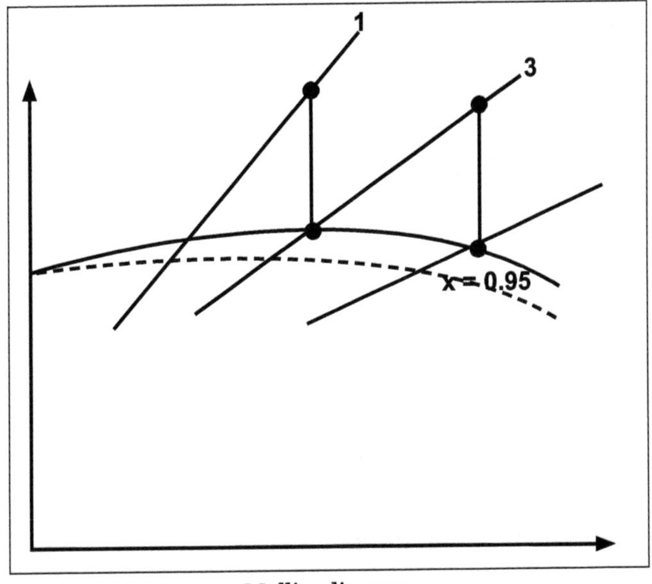

Mollier diagram.

$P_2 = P_3 = 13.5$ bar
$= 1.35$ MPa

$$\eta_1 = \frac{(h_2 - h_2) + (h_3 - h_4) - W_P}{(h_1 - h_6) + (h_3 - h_2)} = 43.6\%$$

$$M = \frac{1}{W_T - W_P} = 2.05 \text{ kg / kWh}$$

6. Let us consider a right bank of 0.03m3 capacity contains vapour at 80 k Pa. If the wet vapour mass is 12 kg, calculate the heat added and the quality of the mixtures when the pressure inside the tank reaches 7 MPa.

Solution:

Given:

$V_1 = V_2 = 0.03\ m^3$;

$P_1 = 80$ kPa m = 12 kg ; $P_2 = 7$ MPa

$$V_1 = \frac{V_1}{m} = 0.03/12 = 0.0025 m^3/kg$$

From steam tables, corresponding to 80 kpa, read V_f, V_g, h_f and h_{fg}

$h_f = 391.7$ kJ/ kg;

$h_{fg} = 2274.21$ kJ/kg

$V_g = 2.0869\ m^3$ / kg

$V_f = 0.001039\ m^3$ / kg

Specific volume, $V_1 = V_f + x_1 V_{fg}$

$$0.0025 = 0.001039 + x_1 \times [2.0869 - 0.001039]$$

$$x_1 = 0.007 (\because V_{fg} = V_g - V_f)$$
$$h_1 = h_f + x_1;\ hfg = 391.7 + 0.007 \times 2274.21$$
$$= 407.62\ kJ/kg$$

For rigid vessel,

$$V_1 = V_2\ (\therefore V_2 = 0.0025\ m^3/kg)$$

From steam tables, corresponding to 7 MPa, read

$$h_f = 1317.2\ kJ/kg\ ; h_{fg} = 1442.7\ kJ/kg,$$
$$V_g = 0.027368\ m^3/kg; V_f = 0.001035/m^3/kg$$

Here, $V_{g2} > V_{21}$.

The steam is in wet condition

$$V_2 = V_f + x_2 V_g \Rightarrow 0.0025$$
$$= x_2 \times (0.027368 - 0.0010351)$$
$$x_2 = 0.04$$

$$\therefore h_2 = hf_2 + x_2 hf_g = 1317.2 + 0.044 \times 1442.7 = 1380.68 \text{ kJ/kg}$$

By first law of thermodynamics,

$$Q = W + \Delta U$$

For rigid vessel,

$$V_1 = V_2$$

So work transfer

$$W = 0 \ (\therefore Q = \Delta U)$$

Total heat transfer

$$= m[u_2 - u_1] = m[[h_2 - P_2V_2] - [h_1 - P_1V_1]$$

$$= m\,[(h_2 - h_1) - V_1(P_2 - P_1)]\ [\because V1 = V2]$$

$$\text{Total heat transfer, } Q = 12[(1380.68 - 407.62) - 0.0025(7000 - 80)]$$

$$= 11469.12 \text{ kJ}$$

7. Let us consider 1 kg of steam initially dry saturated at 1.1 MPa expands in a cylinder following the law $PV^{1.13} = C$. The pressure at the end of expansion is 0.1 MPa. Determine (i) the final volume, (ii) Final Dryness Fraction, (iii) Work done, (iv) The change in internal energy, (v) The heat transferred.

Solution:

Given:

$$m = 1\text{kg},\ P_1 = 1.1 \text{ MPa},\ P_2 = 0.1\text{MPa},\ Pv^{1.13} = C.$$

From steam table at 1.1 MPa

$V_1 = V_{g1} = 0.17739 \text{ m}^3/\text{kg}$

$h_2 = h_{g2} = 2779.7 \text{ kJ/kg}$

From Polytropic relation,

$$P_1V_1^{1.13} = P_2V_1^{1.13}\left[P_1V_1^{n} = P_1V_2^{n}\right]$$

$$V_2 = V_1 \times \left(\frac{P_1}{P_2}\right)^{1/1.13}$$

$$=0.167739 \times \left(\frac{1.1}{0.1}\right)^{1/1-13}$$

$$=1.48 \text{ m}^3$$

We know that,

$$V_2 = V_{f2} + x_2 V_{fg2}$$

$$1.48 = 0.0001043 + x_2 x\ 1.6928$$
$$X_2 = 0.874$$

Work done,

$$W = \frac{P_1V_1 - P_2V_2}{n-1} = \frac{1100 \times 0.17739 - 100 \times 1.48}{1.13-1}$$

$$W = 362.53 \text{ kJ}$$

We know that

$$h_2 = h_{f2} + x_2 h_{fg2}$$

$$h_2 = 417.5 + 0.874(2257.9)$$
$$h_2 = 2390.9 \text{ kJ/kg}$$

Change in internal energy,

$$\Delta U = u_1 - u_2$$

$$= (h_1 - P_1V_1) - (h_2 - P_2V_2)$$
$$= (h_1 - h_2) - (P_1V_1 - P_2V_2)$$
$$= (27779.7 - 2390{:}9) - (1100 \times 0.17739 - 100 \times 1.48)$$
$$= 341.671 \text{ kJ}$$

By 1st law of thermodynamics,

$$Q = W + \Delta U$$

$$Q = 362.53 + 341.671 = 704.2 \text{ kJ}$$

3.3 Boiler and its Classification

A steam boiler or generator is a closed vessel made of steel whose function is to transfer heat produced by the combustion of fuel to water to produce steam.

The steam so produced may be supplied:

- To a steam engine or steam turbine.
- For industrial process work in sugar factories, cotton mills etc.
- For heating installations.

Classification of boilers

The steam boilers are classified as:

- According to flow of water and hot gases:
    - Fire Tube Boilers.
    - Water Tube Boilers.
- According to the method of firing:
    - Internally fired boilers.
    - Externally fired boilers.
- According to the Pressure developed:
    - Low pressure boilers.
    - High pressure boilers.

a) According to Flow of Water and Hot Gases

Fire Tube Boilers

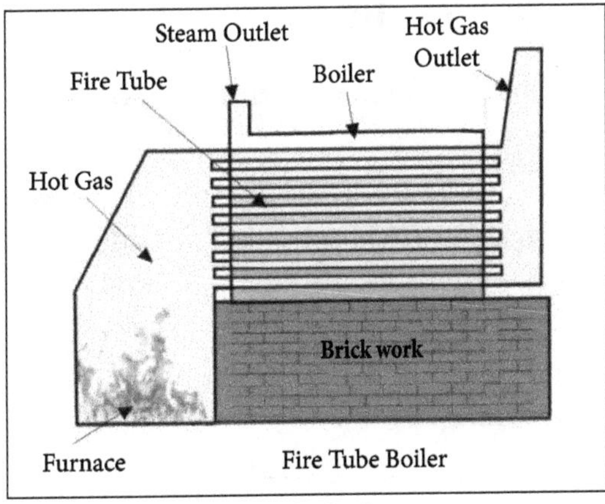

Fire tube boiler.

- The hot gases pass through the tubes surrounded by water.
- The water is get heated up and converted into steam.
- The exhaust gases are sent to atmosphere through chimney.

E.g.: Locomotive boiler, Lancashire boiler.

Water Tube Boilers

Water is circulated through number of tubes and the hot flue gases flow over these tubes. A number of tubes are connected with boiler drum through headers. The hot gases flow over these tubes many times before escaping through the stack. The water is converted into steam and steam occupies steam space.

E.g.: Babcock & Wilcox, Stirling, BHEL boiler, Velox, Lamont, Loeffler boilers.

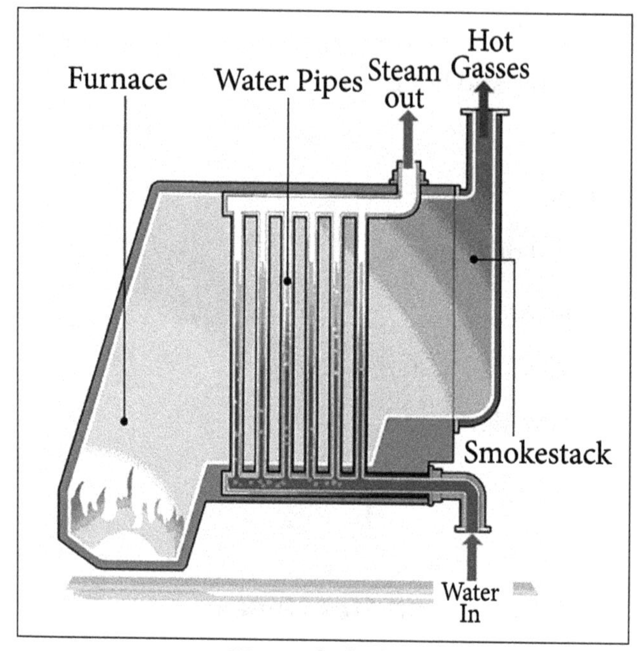

Water tube boiler.

b) According to the Method of Firing

In internally fired boilers, the furnace grate is provided inside the boiler shell.

E.g.: Lancashire, Locomotive boilers.

In externally fired boilers, the furnace grate is provided outside or built under the boiler shell.

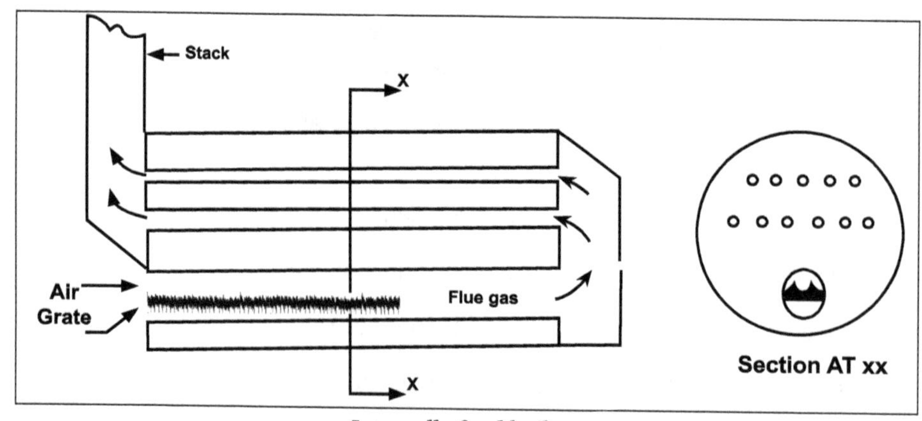

Internally fired boiler.

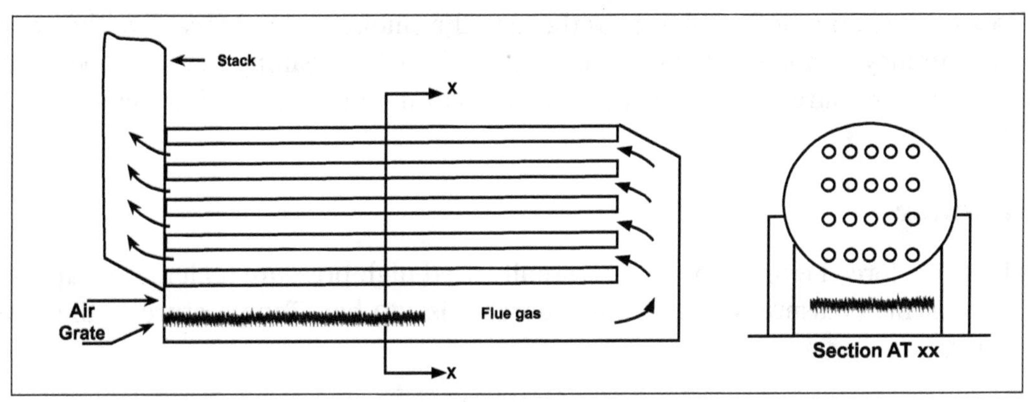

Externally fired boiler.

c) According to the Pressure Developed

In Low Pressure Boilers, Steam is produced at a pressure lower than 80 bar. (E.g. Cochran, Lancashire, Locomotive). In High Pressure Boilers, Steam is produced at a pressure more than 80 bar.

E.g: Lamont, Velox, Benson, Loeffler boiler

Cochran Boiler

Coal is fed into the grate through the fire hole and burnt. Ash formed during the burning is collected in the ash pit provided just below the grate. Ash is then removed manually. The hot gases from the grate pass through the combustion chamber to the horizontal fire tubes and transfer the heat by convection.

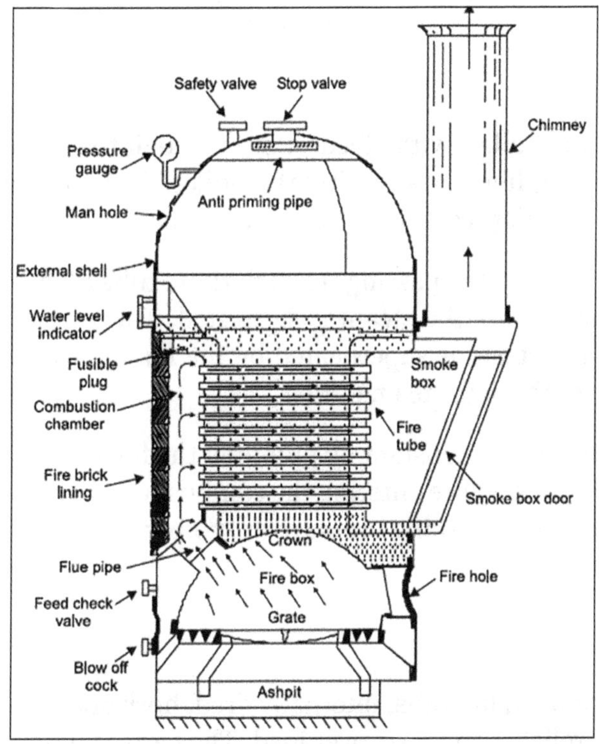

Cochran Boiler.

The flue gases coming out of fire tubes pass through the smoke box and escape to the atmosphere through the chimney. Smoke box is provided with a door for cleaning the fire tubes and smoke box. The working pressure and steam capacity of Cochran boiler are 6.5 bar and 3500 kg/hr respectively.

Lamont Boiler

It is a water tube, forced circulation and externally fired high pressure boiler. The capacity of the plant is 50 tons/hr. Pressure of the steam generated is 170 bar. Temperature of the steam produced is 500°C.

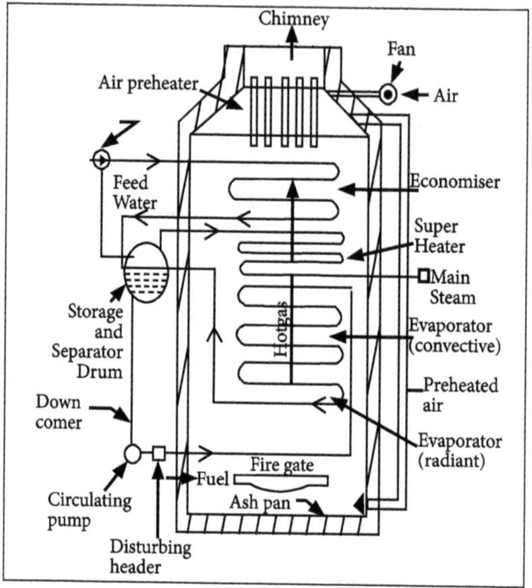

Lamont Boiler.

Working

Feed water is pumped to the boiler by the feed pump through the economizer. Economizer preheats the feed water by using hot gases leaving the boiler. The circulating pump circulates the water from the drum under high pressure to prevent the tubes from being overheated.

Water is evaporated into steam when passing through these tubes. The water and steam from the tube enters the boiler drum where the steam is separated. This steam is passed through a convection super heater and the steam is superheated by the flue gases. This super-heated steam is supplied to the prime mover through steam outlet.

The water level in the drum is kept constant by pumping the feed water into the boiler drum. The air is preheated by the flue gases before entering the combustion chamber to aid the combustion of the fuel. This type of boiler has a working pressure of 170 bar. They can produce the steam at the rate of 45000 kg per hour.

Lancashire Boiler

Lancashire Boiler is a stationary, fire tube, internally fired, horizontal and natural circulation boiler. Lancashire boilers are reliable and bear over load. They are economical within their operating capacity. However, they are bulky and initially raise steam very slowly. Working pressure in Lancashire boilers are in the range of 0.7 MPa to 2 MPa and efficiency of the boiler is about 65%–70%. Size of this boiler depends upon size of shell which may be 2m to 3m in diameter and 6m to 10m in length.

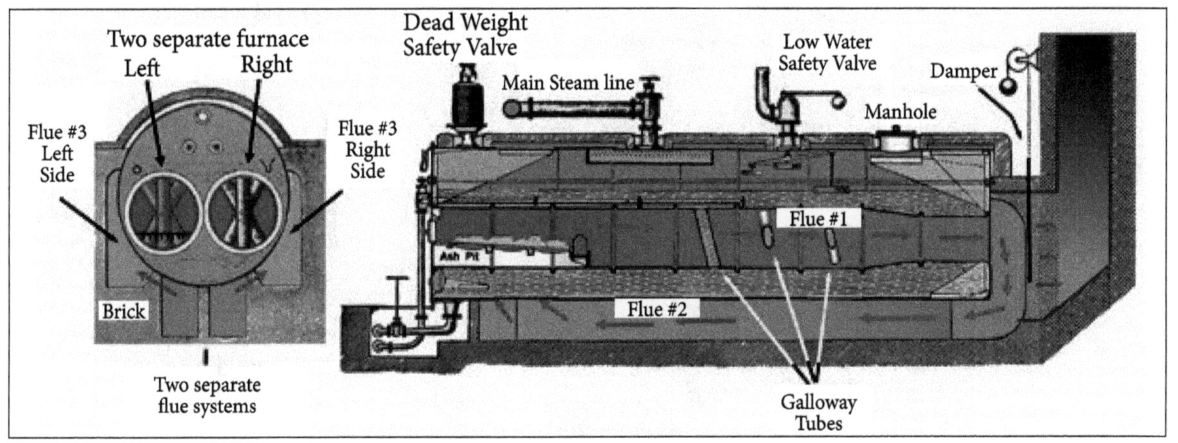

Lancashire boiler.

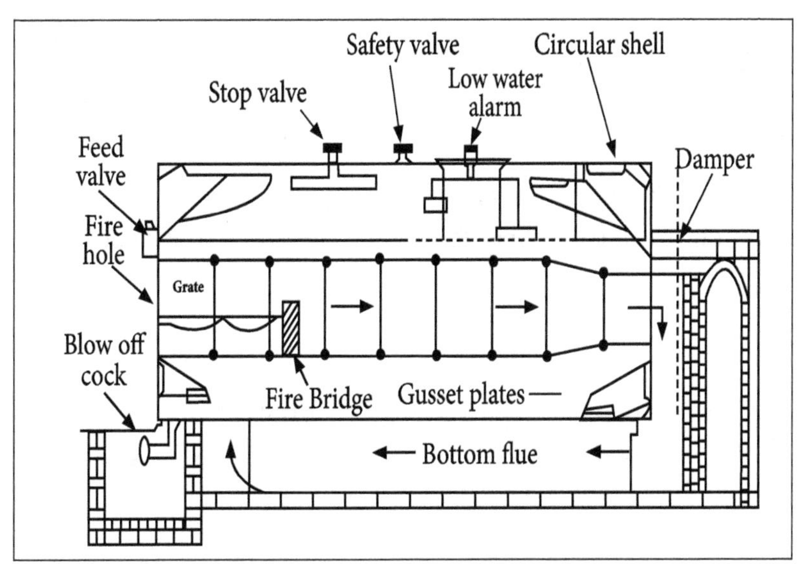

Cross section view.

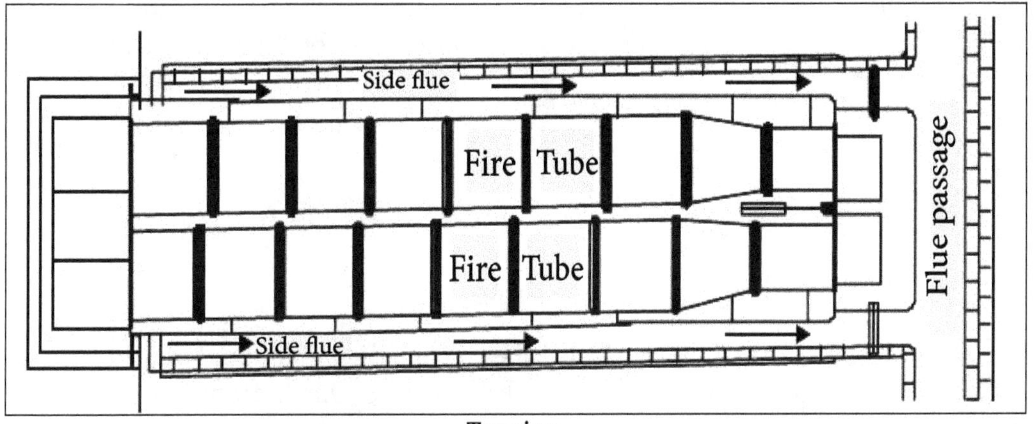

Top view.

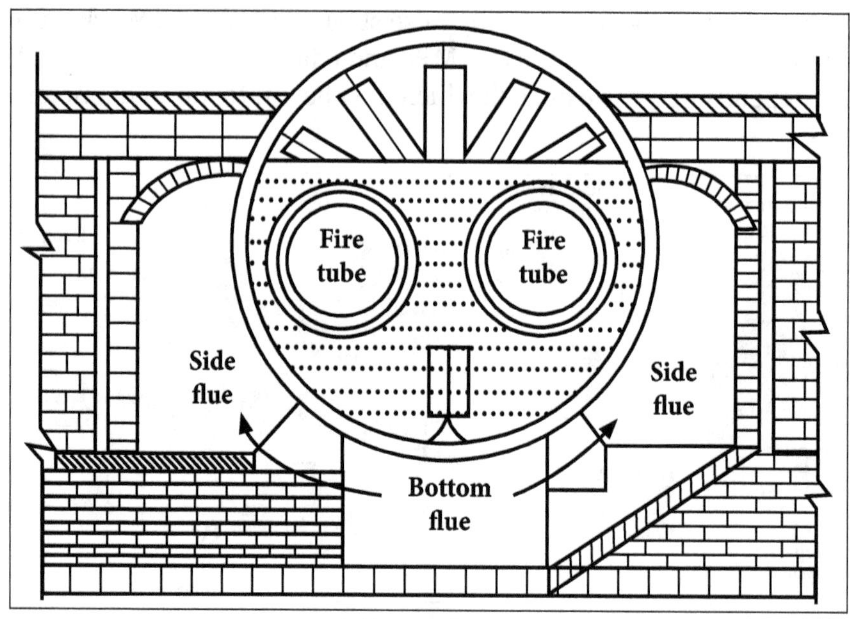

Gusset plate.

Construction of Lancashire Boiler

Cylindrical Shell

Lancashire Boiler consists of a horizontal cylindrical shell of 1.75m to 2.75m diameter. Its length varies from 7.25m to 9m. It is partly filled up with water. The water level inside the shell is above the furnace tubes.

Dampers

Dampers are fitted at the end of side flues to control the draught and thus regulate the rate of generation of steam. These dampers are operated by chain passing over a pulley on the front of the boiler.

Fire Bridge

Fire Bridge is provided to prevent fuel from falling over the end of furnace. Fire Bridge also helps in producing a better mixture of air and gases for perfect combustion by partly enveloping the combustion space.

Grate

The fire grate is provided at one end of the flue tubes on which solid fuel is burnt.

Furnace tubes, bottom flue and side flues

Lancashire Boiler consists of two internal flue tubes having diameter about 0.4 times that of shell. The flues are built-up of ordinary brick lined with fire bricks. One bottom flue and two side flues are formed by brick setting.

Blow off Cock

It is used for removing mud etc. that settles down at the bottom of the boiler, by forcing out some of the water. It is also used to empty water in the boiler.

Working of Lancashire Boiler

In Lancashire Boiler, first the coal is fed on the grate where combustion of coal takes place. The Flue gases pass through the internal flue tubes and then pass downwards. The flue gases travels through the bottom flue to the front of the boiler. Then they are split up into two steams and flow through the side flues to the rear end of the boiler. The flue gases then pass through the dampers into the main flue and are finally discharged through the chimney into the atmosphere. The heat of flue gases is transferred to the water to generate steam.

Babcock and Wilcox Boiler

The Babcock and Wilcox Boiler consist of:

- Steam and water drum (boiler shell).
- Water tubes.
- Uptake-header and down corner.
- Grate.
- Furnace.
- Baffles.
- Super heater.
- Mud box.
- Inspection door.
- Damper.

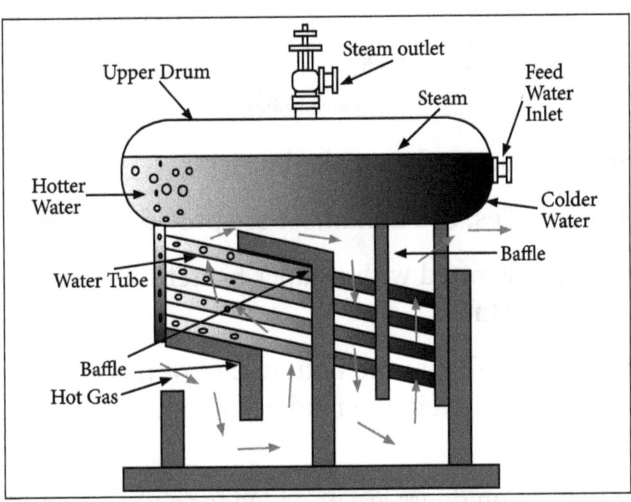

Wilcox Boiler.

Steam and water drum (boiler shell)

One half of the drum which is horizontal is filled up with water and steam remains on the other half. It is about 8 meters in length and 2 meter in diameter.

Water tubes

- Water tubes are placed between the drum and furnace in an inclined position to promote water circulation. These tubes are connected to the uptake-header and the down-comer as shown.
- Uptake-header and down-corner (or down take-header).
- The drum is connected at one end to the uptake-header by short tubes and at the other end to the down-corner by long tubes.

Grate: Coal is fed to the grate through the fire door.

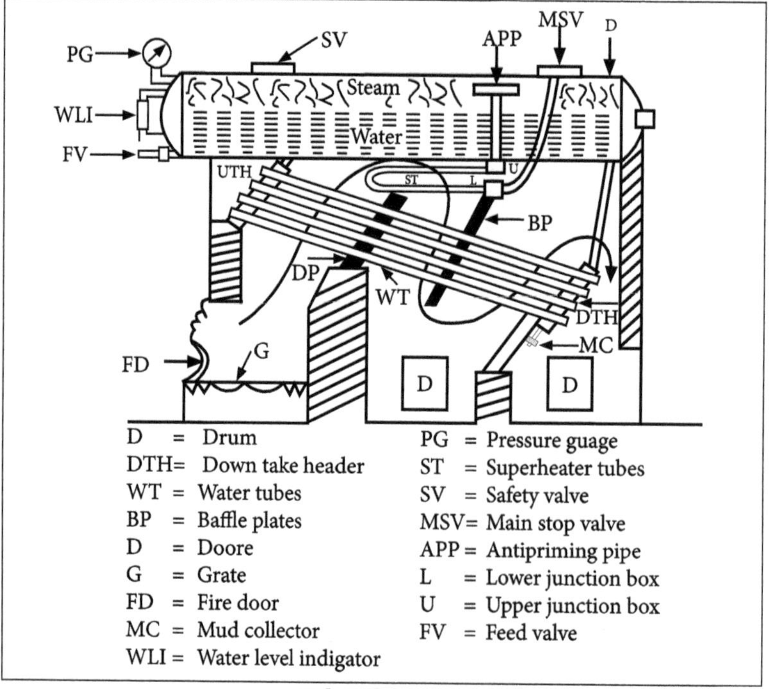

Babcock boiler.

- Furnace: Furnace is kept below the uptake-header.
- Baffles: The fire-brick baffles, two in number are provided to deflect the hot flue gases.
- Super heater: The boiler is fitted with a super heater tube which is placed just under the drum and above the water tubes.
- Mud box: Mud box is provided at the bottom end of the down comer. The mud or sediments in the water are collected in the mud box and it is blown-off time to time by means of a blow off cock.
- Inspection doors: Inspection doors are provided for cleaning and inspection of the boiler.

Working Babcock and Wilcox Boiler

Coal is fed to the grate through the fire door and is burnt.

Flow of Flue Gases

The hot flue gases rise upward and pass across the left side portion of the water tubes. The baffles deflect the flue gases and hence the flue gases travel in the zigzag manner over the water tubes and along the super heater. The flue gases finally escape to atmosphere through chimney.

Water Circulation

That portion of water tubes which is just above the furnace is heated comparatively at a higher temperature than the rest of it. Water density being decreased, rises into the drum through the up-take header. Here the steam and water are separated in the drum. Steam being lighter is collected in the upper part of the drum. The water from the drum comes down through the down comer into the water tubes.

A continuous circulation of water from the drum to the water tubes and water tubes to the drum is thus maintained. The circulation of water is maintained by convective currents and is known as "natural circulation". A damper is fitted as shown to regulate the flue gas outlet and hence the draught.

The boiler is fitted with necessary mountings. Pressure gauge and water level indicator are mount-ed on the boiler at its left end. Steam safety valve and stop valve are mounted on the top of the drum. Blow off cock is provided for the periodical removed of mud and sediments collected in the mud box.

Salient features of Babcock and Wilcox Boiler:

- Its overall efficiency is higher than a fire tube boiler.
- The defective tubes can be replaced easily.
- All the components are accessible for inspection even during the operation.
- The draught loss is minimum compared with other boiler.
- Steam generation capacity and operating pressure are high compared with other boilers.
- The boiler rests over a steel structure independent of brick work so that the boiler may expand or contract freely.
- The water tubes are kept inclined at an angle of 10 to 15 degree to promote water circula-tion.

Advantages water tube boilers over fire tube boilers:

- Steam can be generated at very high pressures.
- Heating surface is more in comparison with the space occupied, in the case of water tube boilers.

- Steam can be raised more quickly than is possible with a fire tube boiler of large water capacity. Hence, it can be more easily used for variation of load.
- The hot gases flow almost at right angles to the direction of water flow. Hence maximum amount of heat is transferred to water.
- A good and rapid circulation of water can be made.
- Bursting of one or two tubes does not affect the boiler very much with regard to its working. Hence water tube boilers are sometimes called as safety boilers.
- The different parts of a water tube boiler can be separated. Hence it is easier to transport.
- It is suitable for use in steam power plants.

Disadvantages of water tube boilers over fire tube boilers:

- It is less suitable for impure and sedimentary water, as a small deposit of scale may cause the overheating and bursting of tubes. Hence, water treatment is very essential for water tube boilers.
- Maintenance cost is high.
- Failure in feed water supply even for a short period is liable to make the boiler overheated. Hence the water level must be watched very carefully during operation of a water tube boiler.

3.3.1 Comparison between water tube boiler and fire tube boiler

S. No.	Water Tube Boiler	Fire Tube Boiler
1.	Water circulates through the tubes and hot flue gases surround the tubes.	Water surrounds the tubes through which hot flue gas passes.
2.	Steam at high pressure can be generated (up to 150 bar)	Pressure is limited to 25 bar.
3.	Since water flows in the tube, direction of water circulation is well defined.	There is no specific direction for water flow.
4.	Steam generation rate is high.	Low rate of steam generation.
5.	Erection and transportation is easy.	Erection and transportation is difficult.
6.	Overall efficiency is high (about 93%).	Low overall efficiency (about 70%).
7.	Can meet the fluctuation in load.	Capable to meet the variation in load over a shorter period.
8.	High operating cost.	Cheap and best.
9.	Used in high capacity power plants.	Suitable only to smaller units.
10.	Chances for accidents are more.	Very rare chances for accidents.

3.4 Boiler Mountings and Accessories

Boiler Mountings

Boiler mountings are primarily intended for the safety of the boiler and for complete control of steam generation process.

Boiler Accessories

Boiler accessories are installed to increase the efficiency of the boiler plants to help in proper working of boiler unit.

Boiler Mountings

- Dead weight safety valve.
- Spring loaded safety valve.
- Fusible plug.
- Pressure gauge.
- Water Level Indicator.

Dead Weight Safety Valve

Weights are placed sufficiently in the weight carrier. The total load on the valve includes the weight of the carrier, the weight of the cover, the weight of the discs and the weight of the valve itself. When the steam pressure exceeds the normal limit, the valve along with the weight carrier is lifted off its seat. Thus, the steam escapes through the discharge pipe.

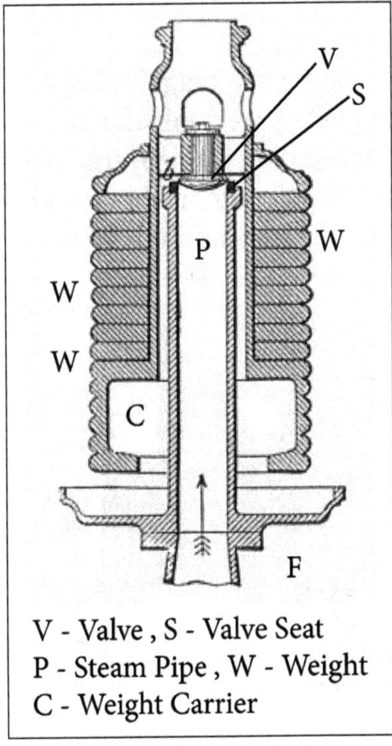

V - Valve , S - Valve Seat
P - Steam Pipe , W - Weight
C - Weight Carrier

Dead Weight Safety Valve.

Spring Loaded Safety Valve

The steam pressure acts below the valves. When the steam pressure is normal the valves are held in their seats tightly by the spring force. When the steam pressure in the boiler exceeds the working pressure, both valves are lifted off their seats. Thus, the steam from the boiler escapes the boiler and steam pressure is reduced. The blow off pressure is adjusted by loosening or screwing the nut.

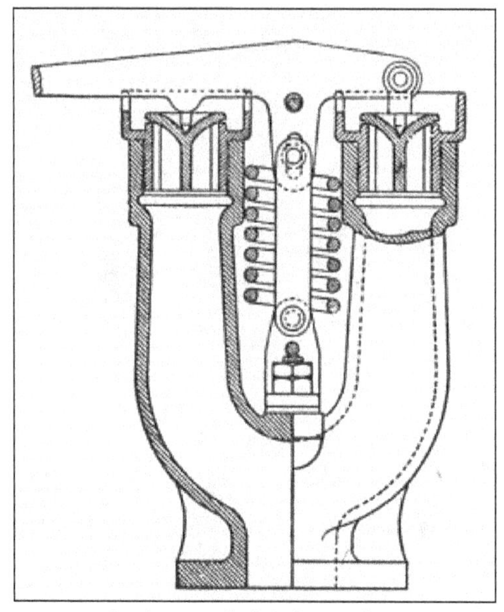

Spring Loaded Safety Valve.

Fusible Plug

Under normal working conditions, the fusible plug is completely covered with water. Hence, the temperature of the plug is not increased appreciably during combustion process. When the water level falls below the safe limit the fusible plug is uncovered from water and exposed to steam. The furnace heat overheats the plug and it melts the fusible metal and copper plug falls down. Due to this, water steam mixture rushes into the furnace and the fire is extinguished.

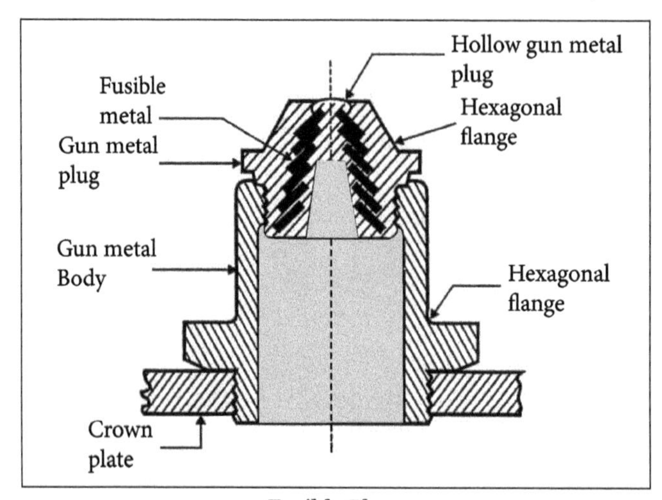

Fusible Plug.

Bourdon Tube Pressure Gauge

The steam pressure is applied to the elliptical cross section of the tube to straighten out slightly. The closed end of the Bourdon tube moves. This movement actuates the toothed sector and pinion rotates. The pointer is mounted on the pinion. Hence, the pointer moves on the graduated dial in clockwise, to indicate the steam pressure.

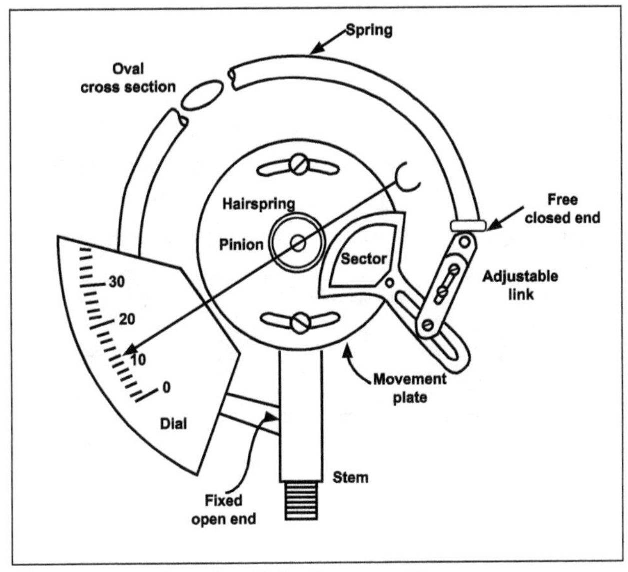

Bourdon Tube Pressure Gauge.

Water Level Indicator

To know the water level in the boiler, the handles of the steam cock and water cock are kept in vertical positions. Water rushes through the bottom casting and steam rushes through the upper casting to the gauge glass tube. The level of water corresponds to the water level in the boiler.

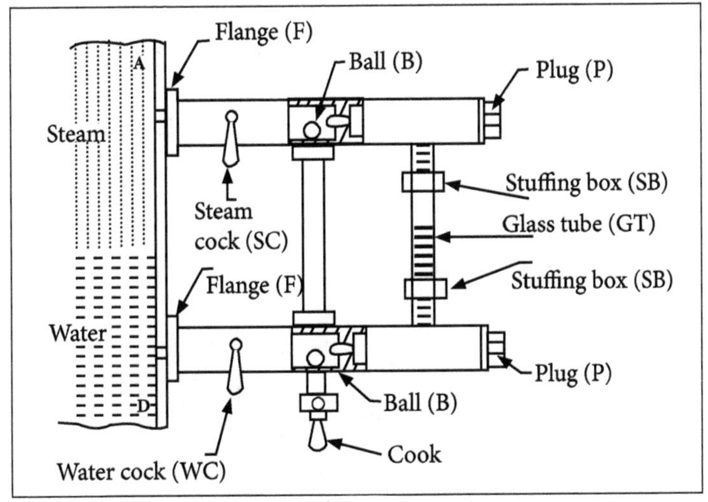

Water Level Indicator.

Boiler Accessories

- Economizer.
- Air Preheater.
- Super Heater.

- Steam Separator.
- Steam Trap.

Economizer

The feed water is pumped to the bottom header and this water is carried to the top header through number of vertical tubes. Hot flue gases are allowed to pass over the external surface of the tubes. The feed water which flows upward in the tubes is heated by the flue gases. This preheated water is supplied to the water. Scrappers are slowly moved up and down to clean the surface of the tubes.

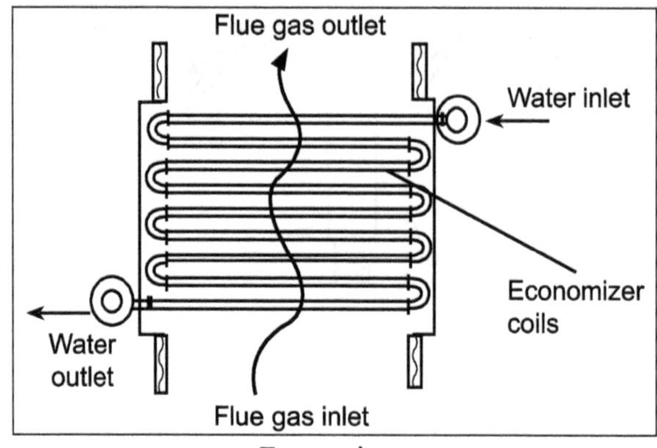

Economizer.

Air Preheater

Hot flue gases pass through the tubes of air preheater after leaving the boiler or economizer. Air and flue gases flow in opposite directions. Baffles are provided in the air preheater and the air passes number of times over the tubes. Heat is absorbed by the air from the flue gases. This preheated air is supplied to the furnace to aid combustion.

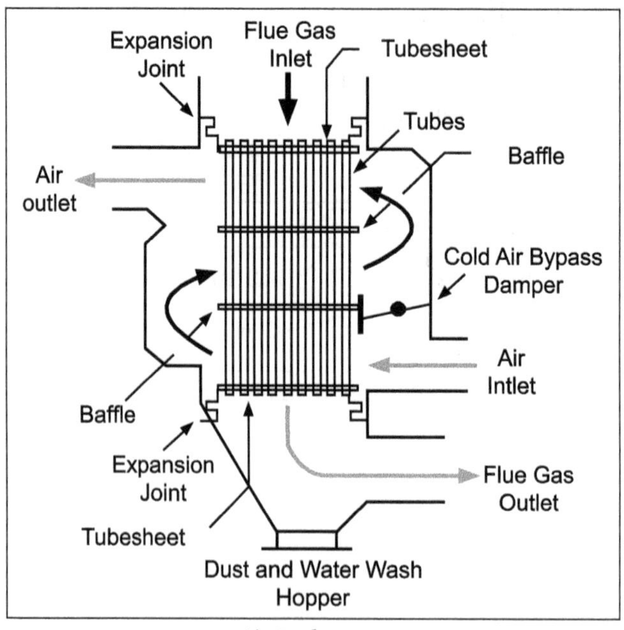

Air Preheater.

Super heater

Steam stop valve is opened. The steam from the evaporator drum is passed through the super heater tubes. First, the steam is passed through the radiant super heater and then to the convective super heater. The steam is heated when it passes through these super heaters and converted into the super-heated steam. This superheated steam is supplied to the turbine through the valve. The steam is allowed into the separator.

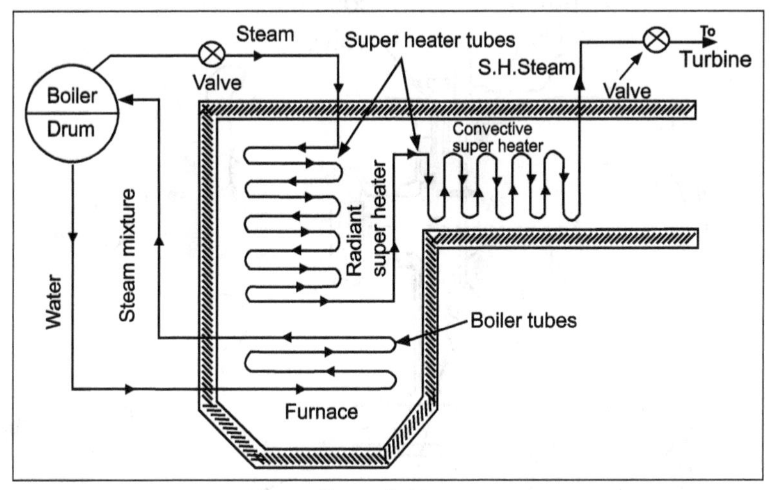

Super heater.

Steam Separator

The steam strikes the baffle plates and the direction of flow is changed. As a result, heavier particles in steam fall down to the bottom of the separator. The separated steam is free from water particles. It is passed to the turbine or engine through the outlet pipe.

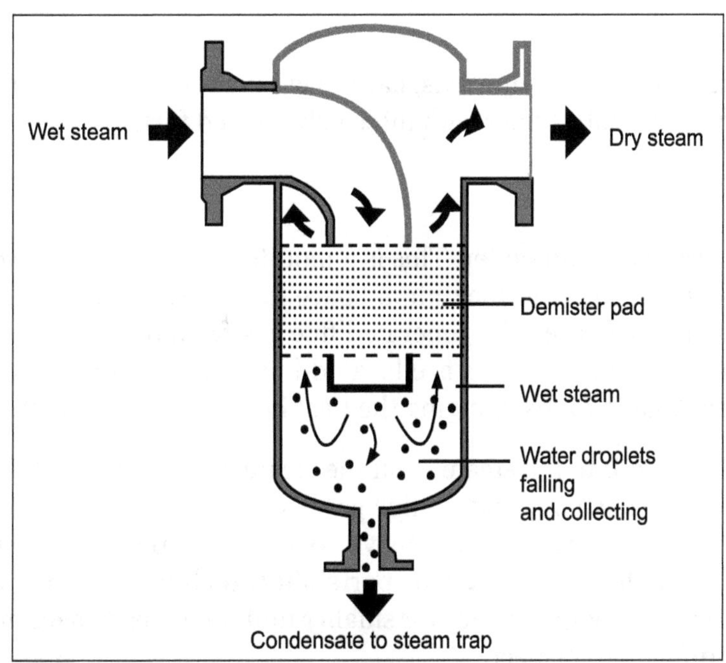

Steam Separator.

Stream Trap

The condensed water enters the steam trap by gravity. When the water level in the trap rises high enough, the ball float is lifted. This causes the valve to open and the water is discharged through the outlet. After the discharge of water, the float moves down. This causes the valve to close again.

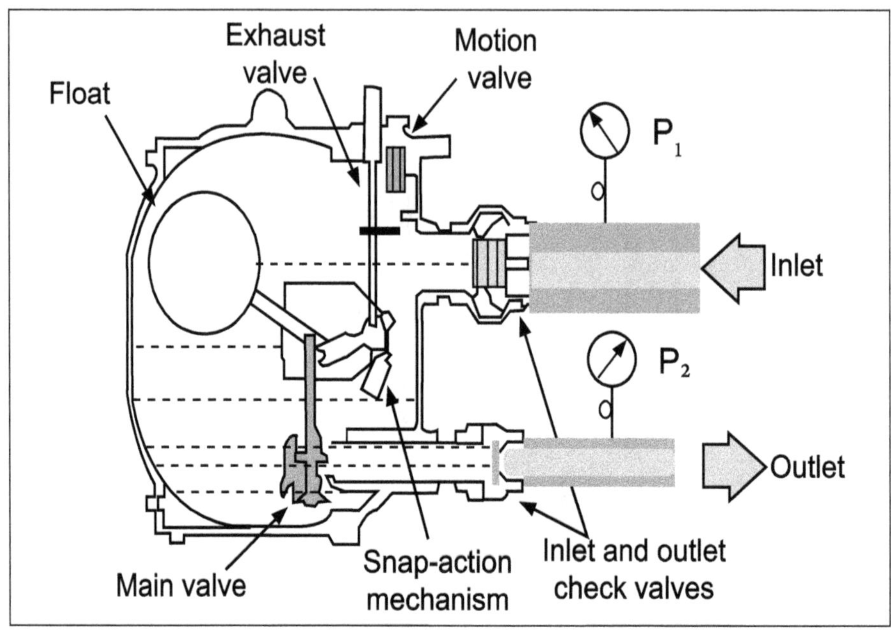

Stream Trap.

3.4.1 Description of Cochran & Bab-cock -Wilcox boiler

Cochran Boiler

Cochran boiler consists of vertical drum axis, natural circulation, natural draft, multi tubular, low pressure, solid fuel fired fire tube boiler with internally located furnace.

Working

The below figure shows a Cochran boiler. This is a modified form of simple vertical boiler which has a hemispherical crown to given maximum space for the steam and very high strength for withstanding high steam pressure. The generated flue gas from the furnace which pass through large number of smaller diameter tubes are located horizontally in the boiler drum. The large heat transfer area is available for exchanging the heat between water and flue gases.

Then the water is converted into the steam from the steam space it is supplied to the plant where the steam is required. The low temperature flue gases will enter the environment through chimney. All the necessary mountings is attached above boiler as per IBR. The advantages of this boiler are its low chimney height, high beaming rate, portability and burning of clay kind of solid as well as liquid fuel whereas it has poor efficiency for smaller unit, high head space, uneconomical in operation and also it is difficult to inspect.

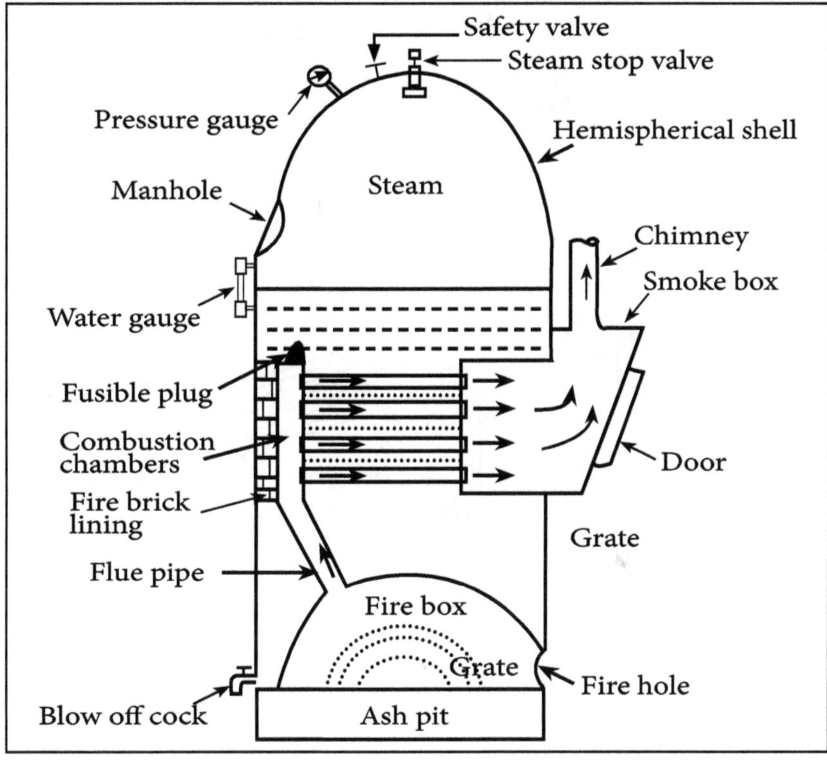

Cochran Boiler.

Babcock and Wilcox Boiler

The Babcock and Wilcox boilers consists of horizontal drum axis, natural circulation, natural draft, multi-tubular, high pressure, stationary, solid fuel fired, water tube boiler with furnace located externally.

Construction and Working

The below figure depicts Babcock and Wilcox boiler. This is high pressure boiler used in power plants. It consists of horizontal boiler drum connected by uptake header and down take header which in turn are connected by number of inclined tubes of water. The flue gases are exchange the heat with the water. The position of baffles cause the gas to move in zigzag way and more heat transfer is possible. A counter flow heating is used.

The draft is regulated by dampers. The water enters the tube through down take header. Due to inclined tubes, the entire tube is not filled with the water. Due to exchange of heat, the steam is separated from the water and through uptake header, it enter the steam space inside the boiler drum. Anti-priming pipe is provided to ensure that only the dry saturated steam enter the super heater via steam stop valve.

It can be built for any width and height because of sectional construction, good circulation, rapid steaming,, safe and free from explosion, fast response to overloads, ease of repair, maintenance and cleaning. It is costlier and fluctuation in water level.

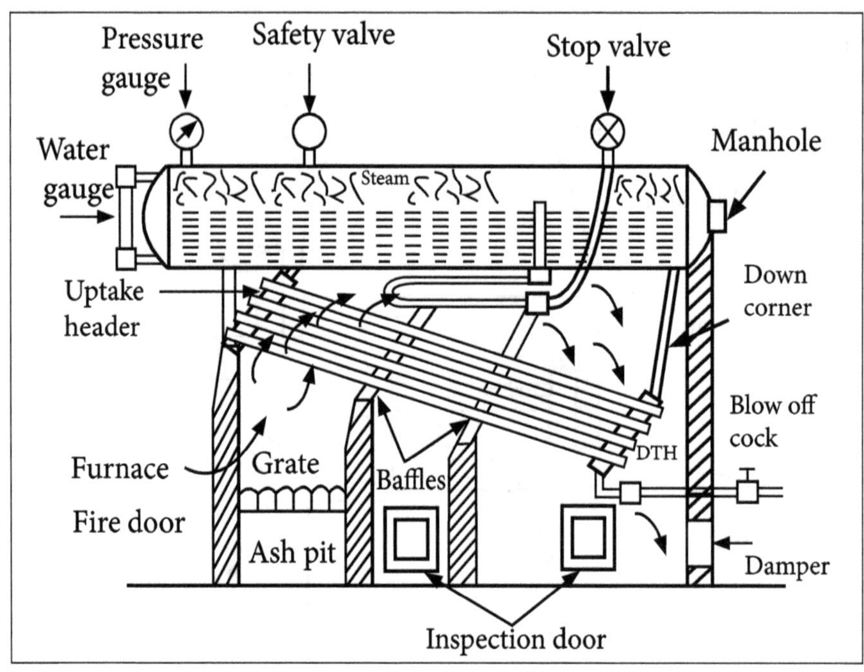

Babcock and Wilcox boiler.

Lancashire Boiler

The Lancashire boilers consists of horizontal drum axis, natural circulation, natural draft, two-tubular, medium pressure, stationary, fire tube boiler with furnace located internally.

Construction and Working

The below figure shows the constructional details of Lancashire boiler along with different boiler mountings, brick work, path of flue gases, furnace etc. Fuel is burnt on the grate and the flue gases can flow from one furnace end to other end of tubes (i.e. from front side to back side of furnace). This is first pass of flue gas through the boiler tubes.

The water is surrounded to the tube. The heat between the water in the boiler drum and the flue gases inside the tube. So the steam is formed. Flue gases available at the backside of the furnace can be diverted in the downward direction due to presence of brick work. (Brick is a very poor conductor of heat energy and can works as insulating material for a given system). So the flue gases can flow from the bottom part of the boiler drum and exchange the heat with water.

This is second pass of flue gases outside the tube. So the flue gases are available at front side. From front, because of brick work, they are divided into two side flues and once again flow backward from the sides of boiler drum and finally are expelled out to stack chimney through main flue. Dampers are provided at the end of side flues to regulate the flow of flue gases.

The disadvantages of the boiler include more floor space, leakage problems through brick-settings, more steaming time, sluggish water circulation, limitation of high pressure of steam and limited space for grate area of furnace. The advantage of Lancashire boiler are large steam space, load fluctuations can easily be met, easy to clean and inspect, reliable, easy to operate and maintain.

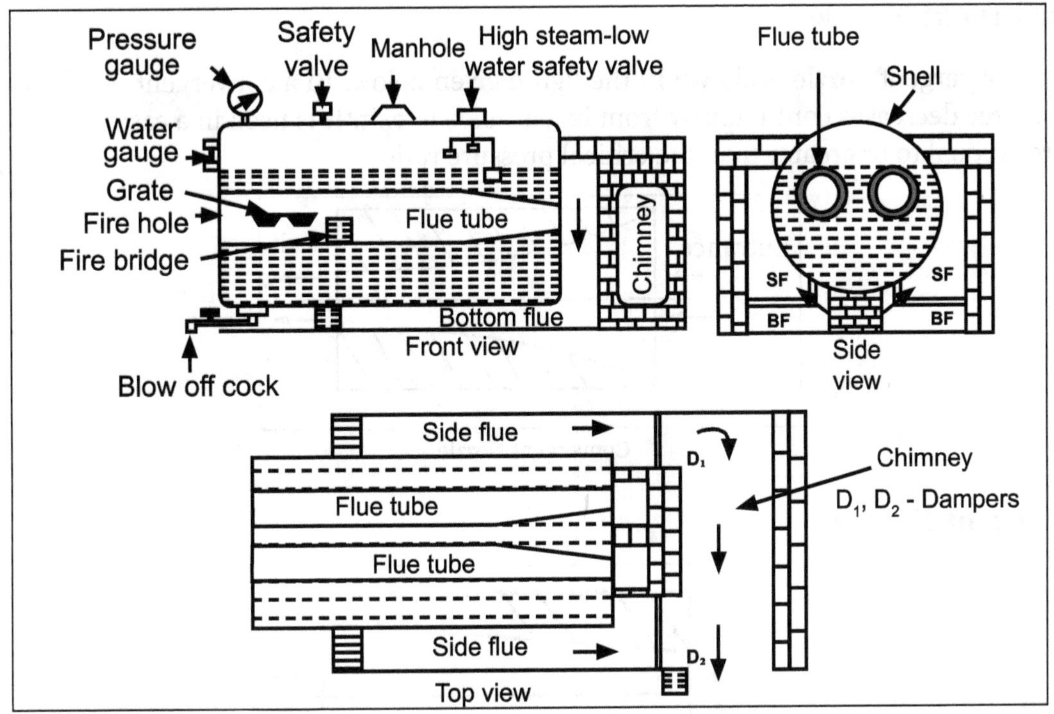

Lancashire Boiler.

3.5 Steam Nozzles: Types of Nozzles

A nozzle is a device of varied cross-sectional area in which the potential energy of steam is converted into the kinetic energy. The increased velocity of steam jet at the exit of the nozzle is due to the decrease in enthalpy of steam.

The nozzles are used in the following engineering applications:

- Steam and gas turbines.
- Jet engines.
- Rocket motors.
- In flow measurement.
- In injectors for pumping feed water into the boiler.
- In injectors for removing air from condensers.
- In water sprinklers.

There are three types of nozzles:

- Convergent Nozzle.
- Divergent Nozzle.
- Convergent-Divergent Nozzle.

1. Convergent Nozzle

A typical convergent nozzle is shown in the figure given below. In a convergent nozzle, the cross sectional area decreases continuously from its entrance to exit. It is used in a case where the back pressure is equal to or greater than the critical pressure ratio.

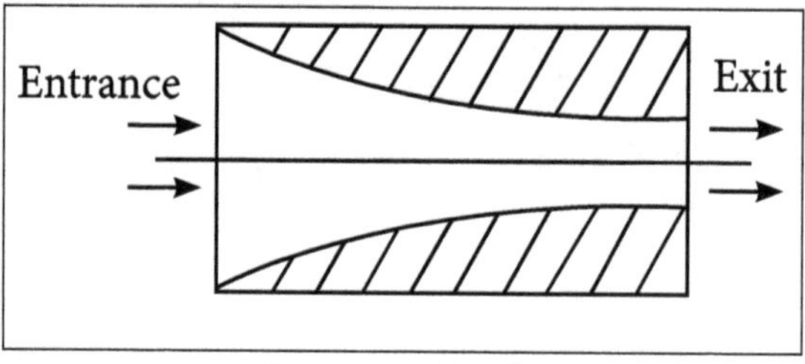

Convergent Nozzle.

2. Divergent Nozzle

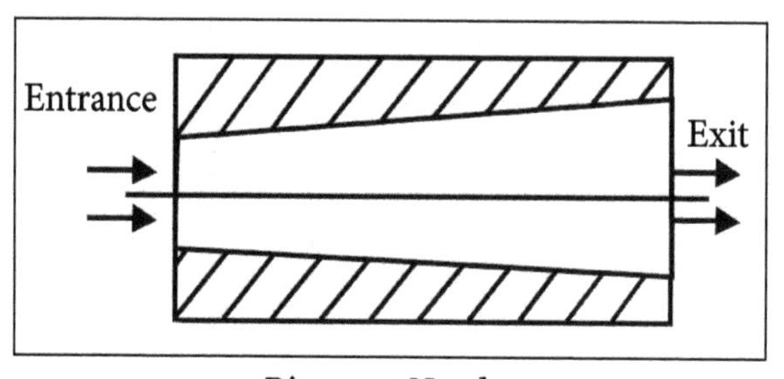

Divergent Nozzle.

The cross sectional area of divergent nozzle increases continuously from its entrance to exit. It is used in a case where the back pressure is less than the critical pressure ratio.

3. Convergent-Divergent Nozzle

In this case, the cross sectional area first decreases from its entrance to throat and then increases from throat to exit. This case is also used in the case where the back pressure is less than the critical pressure. Also, in present day application, it is widely used in many types of steam turbines.

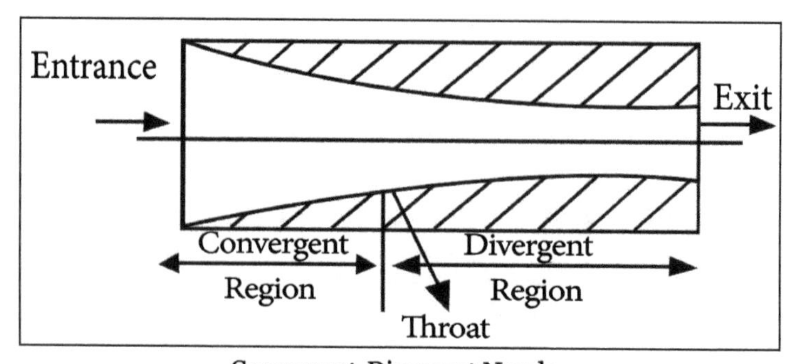

Convergent-Divergent Nozzle.

Suppose a nozzle is used to obtain a supersonic steam starting from low speeds at the inlet. Then the Mach number should increase from Ma=0 near the inlet to Ma>1 at the exit. It is clear that the nozzle must converge in the subsonic portion and diverge in the supersonic portion. Such a nozzle is called a convergent-divergent nozzle.

Flow of Steam Through Nozzles

- The flow of steam through nozzles is known as adiabatic expansion.
- In this process, the steam will have a very high velocity at the end of the expansion and the enthalpy decreases as expansion takes place. The friction exists between the steam and the sides of the nozzle and heat is produced as the result of the resistance to the flow.
- The phenomenon of super saturation occurs in the flow of steam through nozzles because of the time lag in the condensation of the steam during the expansion.

Thermodynamic Analysis

When the steam flows through a nozzle, there is some loss in its enthalpy or total heat which takes place due to the friction between the nozzle surface and the flowing steam.

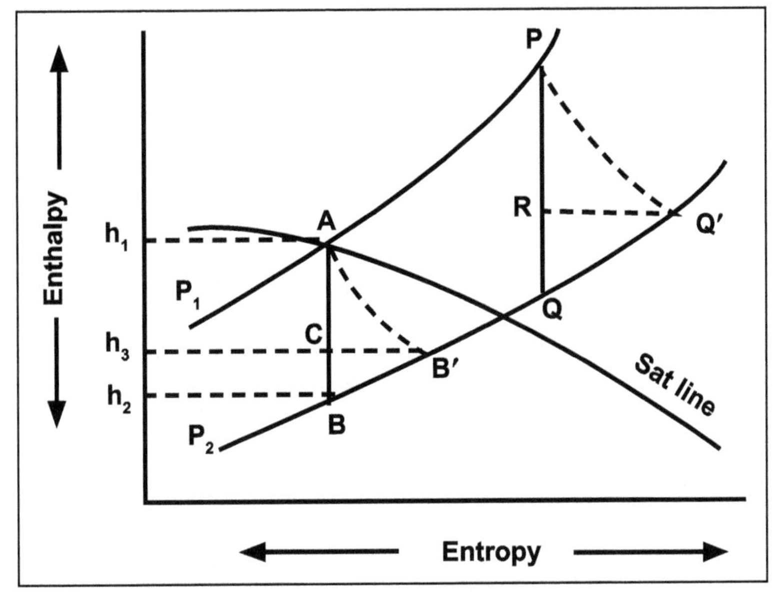

Thermodynamic analysis.

- First, locate the point A for the initial conditions of the steam. It is a point, where the saturation line meets the initial pressure (P_1) line.
- Now draw a vertical line through A to meet the final pressure (P_2) line. As the flow through the nozzle is isentropic, that is expressed by a vertical line AB. The heat drop ($h_1 - h_2$) is known as isentropic heat drop.
- Because of friction in the nozzle the actual heat drop in the steam will be less than ($h_1 - h_2$). Let this heat drop be shown as AC instead of AB.
- As the expansion of steam ends at the pressure P_2, therefore final condition of steam is obtained by drawing a horizontal line through C to meet the final pressure (P_2) Line at B'.
- Finally the actual expansion of steam in the nozzle is expressed by the curve AB' (adiabatic expansion) instead of AB (isentropic expansion). Then the actual heat drop ($h_1 - h_3$) is known as useful heat drop.

Assumptions

- Steady state.
- Adiabatic boundaries (i.e. the fluid Velocity is very high, so there is no time available for heat exchange with surroundings.).
- Mass average velocities adequate for calculations.
- No shaft work.
- Equilibrium states at inlet and outlet.
- Change in potential energy is negligible.

3.5.1 Isentropic Flow through Nozzles

Nozzle

A nozzle is a device which is used to accelerate the fluid. During this process, velocity of fluid increases with decreasing pressure.

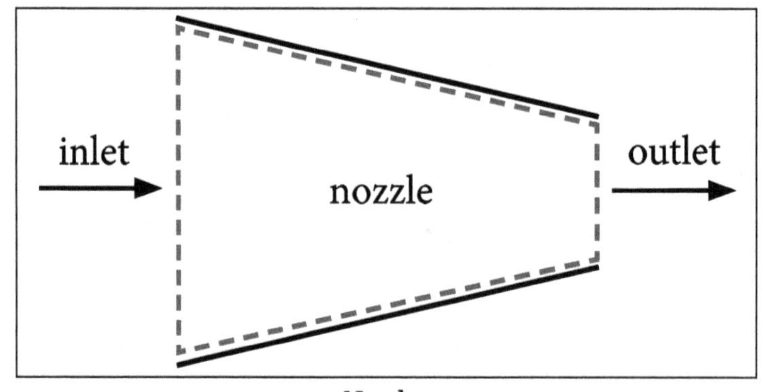

Nozzle.

The flow of steam through a nozzle is isentropic which can be represented by an equation,

$$pv^n = \text{Constant},$$

Where,

n is the index of expansion.

n = 1.135 for saturated steam

n = 1.3 for superheated steam in the expansion through the nozzle, gain in kinetic energy is achieved due to loss in heat. By neglecting initial velocity (approaching velocity), the gain in kinetic energy is given by, $V_2^2/2$.

Drop in heat energy during isentropic process,

$$= \frac{n}{n-1}(p_1v_1 - p_2v_2)$$

Since gain in K.E is equal to drop in heat energy.

$$\frac{V_2^2}{2} = \frac{n}{n-1}(p_1v_1 - p_2v_2)$$

$$= \frac{n}{n-1}p_1v_1\left(1 - \frac{p_2}{p_1}\cdot\frac{v_2}{v_1}\right)$$

From p.v.t relationship,

$$\frac{v_2}{v_1} = \left(\frac{p_1}{p_2}\right)^{\frac{1}{n}} = \left(\frac{p_2}{p_1}\right)^{\frac{-1}{n}}$$

$$v_2 \quad v_1\left(\text{—}\right)^{-} \quad \text{(i)}$$

$$\frac{v_2^2}{2} = \frac{n}{n-1}p_1v_1\left(1 - \frac{p_2}{p_1}\left(\frac{p_2}{p_1}\right)^{\frac{-1}{n}}\right)$$

$$= \frac{n}{n-1} \text{ x } p_1v_1\left[1 - \left(\frac{p_2}{p_1}\right)^{1-\frac{1}{n}}\right]$$

$$v_2 = \sqrt{\frac{2n}{n-1}p_1v_1\left[1 - \left(\frac{p_2}{p_1}\right)^{\frac{n-1}{n}}\right]} \qquad \text{...(ii)}$$

From continuity equation, we know that mass of steam discharged from the nozzle per second is given by,

$$m = \rho A V_2 \qquad \text{...(iii)}$$

Where,

ρ-Density of steam A is the cross sectional area

V_2 -Exit velocity. Also, w.k.t. density

$$= \frac{1}{\text{Specific volume}} = \frac{1}{v_2} \qquad \text{...(iv)}$$

Sup (ii) and (iv) in equation (iii) given above

$$m = \frac{AV_2}{v_2} = \frac{A}{v_2}\sqrt{\frac{2n}{n-1}p_1v_1\left[1 - \left(\frac{p_2}{p_1}\right)^{\frac{n-1}{n}}\right]}$$

Therefore,

From p.v.t relationship we know that sub (i) in above equation,

$$m = \frac{A}{v_1\left(\frac{p_2}{p_1}\right)^{\frac{n-1}{n}}}\sqrt{\frac{2n}{n-1}p_1v_1\left[1-\left(\frac{p_2}{p_1}\right)^{\frac{n-1}{n}}\right]}$$

$$= \frac{A}{v_1}\left(\frac{p_2}{p_1}\right)^{\frac{n-1}{n}}\sqrt{\frac{2n}{n-1}p_1v_1\left[1-\left(\frac{p_2}{p_1}\right)^{\frac{n-1}{n}}\right]}$$

$$= A\sqrt{\frac{2n}{n-1}\left(\frac{p_2}{p_1}\right)^{\frac{2n}{n}} x \frac{p_1v_1}{v_1^2}\left[1-\left(\frac{p_2}{p_1}\right)^{\frac{n-1}{n}}\right]}$$

$$m = A\sqrt{\frac{2n}{n-1}\cdot\frac{p_1}{v_1}\left[\left(\frac{p_2}{p_1}\right)^{\frac{2}{n}}-\left(\frac{p_2}{p_1}\right)^{\frac{n+1}{n}}\right]}$$

Coefficient of Friction in Nozzle

The coefficient of friction of nozzle or nozzle efficiency is defined as the ratio of actual enthalpy drop to the isentropic enthalpy drop.

$$\text{Coefficient of friction} = \frac{\text{Actual enthalpy drop}}{\text{Isentropic emthalpy drop}}$$

3.5.2 Effect of Friction on Nozzle Efficiency

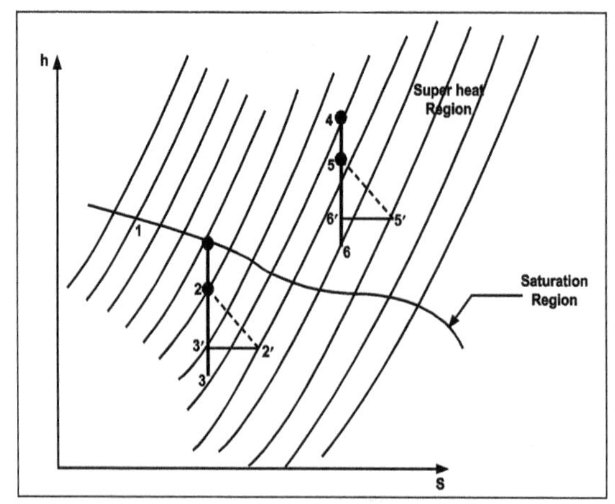

Effect of friction on steam flow through a nozzle.

The point 1 represents the initial condition of steam which enters the nozzle in a dry saturated state. If the friction is neglected, the expansion of steam from entry to throat is represented by the vertical line 1-2 and that from the throat to the exit by 2-3. Now if the friction were taken into account the heat drop would have been somewhat less than 1-3.

Let this heat drop be 1 - 3'. From 3' draw a horizontal line which cuts the same pressure line on which point 3 lies, at the point 2' which represents the final condition steam. It may be noted that dryness fraction of steam is more at point 2 than at point 3. Hence the effect of friction is to improve the quality of steam.

The value of coefficient 'k' in the equation for the velocity of expanding steam is given by,

$$k = \frac{\text{Actual heat drop}}{\text{Isentropic heat drop}} = \frac{1-3}{1-3} = \frac{h_1 - h_3}{h_1 - h_3}$$

The actual expansion is represented by the curves 1-2- 2 since the friction occurs mainly between the throat and exit.

On the other words, if the steam at entry to nozzle were superheated corresponding to the point 4, the expansion can be represented by the vertical line 4-6 if friction were neglected and by 4 - 5 - 5′ if the friction were taken into account. In this case $k = 4-6/4-6 = (h_4 - h_{6'})/(h_4 - h_6)$. The point 5 represents the final condition of steam.

It may be noted that the friction tends to superheat steam. Therefore it can be concluded that friction tends to decrease the velocity of steam and increase the final dryness fraction or superheat the steam. The nozzle efficiency is therefore defined as the ratio of actual enthalpy drop to the isentropic enthalpy drop between the same pressures.

$$\text{Nozzle efficiency} = \frac{h_1 - h_3}{h_1 - h_3} \text{ (or) } \frac{h_4 - h_6}{h_4 - h_6}$$

As the case may be; if the actual velocity at exit from the nozzle is C_2 and the velocity at exit when the flow is isentropic is C_3 then using the steady flow energy equation in each case we have,

$$h_1 + \frac{C_1^2}{2} = h_3 + \frac{C_3^2}{2} \text{ (or) } h_1 - h_3 = \frac{C_3^2 - C_1^2}{2},$$

$$\text{And } h_1 + \frac{C_1^2}{2} = h'_2 + \frac{C_{2'}^2}{2} \text{ (or) } h_1 - h'_2 = \frac{C_{2'}^2 - C_1^2}{2}.$$

[In MKS units $\frac{C^2}{2}$ to be represented by $\frac{C^2}{2gj}$]

$$\therefore \text{ Nozzle efficiency} = \frac{C_{2'}^2 - C_1^2}{C_3^2 - C_1^2}$$

When the inlet velocity, C_1 is negligible small then,

$$\text{Nozzle efficiency, } = \frac{C_{2'}^2}{C_3^2} \quad \text{...(i)}$$

Sometimes a velocity co-efficient is defined as the ratio of actual exit velocity to the exit velocity when the flow is isentropic between the same pressures.

$$\text{Velocity Co-efficient} = \frac{C'_2}{C_3}. \quad \text{...(ii)}$$

It can be seen from equations 1 and 2 that the velocity co-efficient is the square root of the nozzle efficiency when the inlet velocity is assumed to be negligible.

Effect of Friction

- It reduces the exit velocity.
- It reduces the efficiency of nozzle.
- The effect of friction is to reduce the available enthalpy drop by 10 to 15%.
- It improves the quality (dryness fraction) of steam.
- Specific volume of steam is increased.
- It increases the entropy.

3.5.3 Critical Pressure Ratio and Maximum Discharge

Critical Pressure Ratio

Critical flow nozzles are also known as sonic chokes. By establishing a shock wave the sonic choke establish a fixed flow rate unaffected by the differential pressure, any fluctuations or changes in downstream pressure. A sonic choke may provide a simple way to regulate a gas flow.

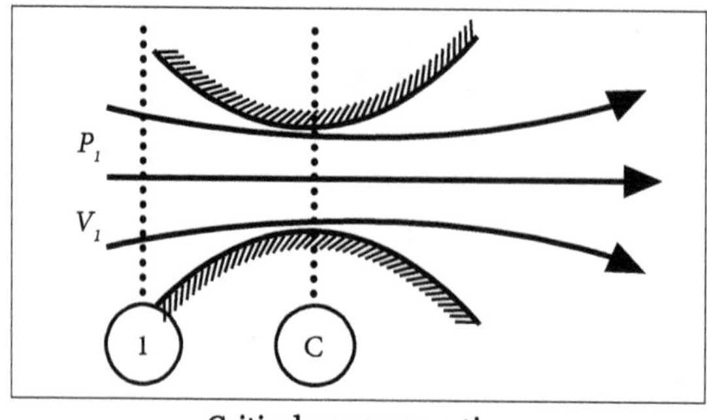

Critical pressure ratio.

Let,

P_1 = Initial pressure of steam in N/m^2.

P_2 = Pressure of steam at throat is N/m^2.

V_1 = Volume of 1 kg of steam at pressure (P_1) in m^3.

V_2 = Volume of 1 kg of steam at pressure (P_2) in m^3.

A = Cross sectional area of nozzle at throat in m^2.

The mass of steam discharged through a nozzle,

$$m = A \times \sqrt{\frac{Zn}{n-1} \times \frac{P_1}{V_1}\left[\left(\frac{P_2}{P_1}\right)^{2/n} - \left(\frac{P_2}{P_1}\right)^{n+1/n}\right]} \quad \text{...(i)}$$

Differentiate the equation (i),

$$\frac{d}{d\left(\frac{P_2}{P_1}\right)}\left[\left(\frac{P_2}{P_1}\right)^{2/n} - \left(\frac{P_2}{P_1}\right)^{n+1/n}\right] = 0$$

$$\frac{2}{n}\left(\frac{P_2}{P_1}\right)^{\frac{2}{n}-1} - \frac{n+1}{n}\left(\frac{P_2}{P_1}\right)^{\frac{n+1}{n}-1} = 0$$

$$\left(\frac{P_2}{P_1}\right)^{\frac{2-n}{n}} \times \left(\frac{P_2}{P_1}\right)^{-1/n} = \frac{n+1}{n} \times \frac{n}{2}$$

$$\left(\frac{P_2}{P_1}\right)^{\frac{2-n}{n}-\frac{1}{n}} = \frac{n+1}{2}$$

$$\left[\frac{P_2}{P_1}\right]^{\frac{1-n}{n}} = \frac{n+1}{2}$$

$$\frac{P_2}{P_1} = \left[\frac{n+1}{2}\right]^{\frac{n}{1-n}} = \left[\frac{n+1}{2}\right]^{\frac{-n}{-(1-n)}}$$

$$\frac{P_2}{P_1} = \left[\frac{2}{n+1}\right]^{\frac{n}{n-1}} \quad \text{...(ii)}$$

The ratio of $\left(\frac{P_2}{P_1}\right)$ is known as critical pressure ratio.

Maximum Discharge Flow through a Nozzle

Co-Efficient of Velocity in Nozzle

Co-efficient of nozzle (or) Nozzle efficiency is defined as ratio of actual enthalpy drop to the isentropic enthalpy drop.

Velocity of Steam flowing through a Nozzle

Consider a unit mass flow of steam through a nozzle.

Let,

V_1 = Velocity of steam at the entrance of nozzle in m/s.

V_2 = Velocity of steam at any section considered in m/s.

h_1 = Enthalpy of total heat of steam entering the nozzle in kJ/kg.

h_2 = Enthalpy of total heat of steam at the section considered in kJ/kg.

We know that for a steady flow process in a nozzle,

$$h1+\frac{01}{1000}\left(\frac{V_1^2}{2}\right)=h_2+\frac{1}{1000}\left(\frac{V_2^2}{2}\right)+\text{Losses} \qquad (\because\ (\phi)15=15)$$

Neglecting losses in a nozzle,

$$\frac{01}{1000}\left(\frac{V_1^2}{2}-\frac{V_2^2}{2}\right)=h_1-h_2$$

$$V_2=\sqrt{200(h_1-h_2)+V_1^2}=\sqrt{V_1^2+200h_d} \qquad \text{...(i)}$$

Where,

h_d = Enthalpy or heat drop during expansion of steam in a nozzle.

$h_d = h_1 - h_2$.

Since the entrance velocity or velocity of approach (V_1) is negligible as compared to V_2, therefore from equation (i),

The relation between the velocity of steam and heat during any part of a steam nozzle is given by.

$$V_2=\sqrt{200h_d}=44.72\sqrt{h_d} \qquad \text{...(ii)}$$

Equation (ii) indicates velocity of steam at exit.

Mass Of Steam Discharged Through a Nozzle

The flow of steam, through the nozzle is entropic, which is approximately represented by the general law,

$$pv^n = \text{constant}.$$

We know that gain in kinetic energy, $= \frac{V_2^2}{2}$.

Heat drop = Work done during Rankine cycle $=\frac{n}{n-1}(P_1V_1-P_2V_2)$.

Since gain in kinetic energy is equal to heat drop, therefore,

$$\frac{V_2^2}{2} = \frac{n}{n-1}(P_1V_1 - P_2V_2)$$

$$= \frac{n}{n-1} x \, P_1V_1\left(1 - \frac{P_2V_2}{P_1V_1}\right) \quad ...(i)$$

We know that $P_1V_1^n = P_2V_2^n$,

$$\frac{V_2}{V_1} = \left(\frac{P_1}{P_2}\right)^{-1/n} = \left(\frac{P_2}{P_1}\right)^{-1/n}$$

$$V_2 = V_1\left(\frac{P_2}{P_1}\right)^{-1/n}$$

$$V_2 = V_1\left(\frac{P_2}{P_1}\right)^{-1/n}$$

Substituting the value of $\mathbf{V_2 / V_1}$ in equation (i),

$$\frac{V_2^2}{2} = \frac{n}{n-1} x \, P_1V_1\left[1 - \frac{P_2}{P_1}\left(\frac{P_2}{P_1}\right)^{-1/n}\right]$$

$$= \frac{n}{n-1} x \, P_1V_1\left[1 - \left(\frac{P_2}{P_1}\right)^{-1/n}\right] \quad ...(ii)$$

$$V_2 = \sqrt{2 \, x \frac{n}{n-1} x \, P_1V_1\left[1 - \left(\frac{P_2}{P_1}\right)^{-1/n}\right]}$$

Mass of steam discharged through nozzle per second,

$$m = \frac{\text{Volume of steam flowing per second}}{\text{Volume of 1 kg of steam at pressure } P_2}$$

$$= \frac{AV_2}{V_2} = \frac{A}{V_2}\sqrt{2 \times \frac{n}{n-1} \times P_1V_1\left[1 - \left(\frac{P_2}{P_1}\right)^{n-1/n}\right]} \quad ...(iii)$$

Substituting the value of V_2 from equation (ii),

$$m = \frac{A}{V_1}\left(\frac{P_1}{P_2}\right)^{-1/n} \sqrt{\frac{2n}{n-1} \times P_1V_1\left[1 - \left(\frac{P_2}{P_1}\right)^{n-1/n}\right]}$$

$$= \frac{A}{V_1}\left(\frac{P_2}{P_1}\right)^{1/n} \sqrt{\frac{2n}{n-1} \times P_1 V_1 \left[1 - \left(\frac{P_2}{P_1}\right)^{n-1/n}\right]}$$

$$= A\sqrt{\left(\frac{P_2}{P_1}\right)^{R/n} \times \frac{2n}{n-1} \times \frac{P_1}{V_1}\left[1 - \left(\frac{P_2}{P_1}\right)^{n-1/n}\right]}$$

$$m = A\sqrt{\frac{2n}{n \quad 1} \times \frac{P_1}{V_1}\left[\left(\frac{P_2}{P_1}\right)^{2/n} - \left(\frac{P_2}{P_1}\right)^{n \ 1/n}\right]} \qquad \text{...(iv)}$$

The equation (iv) gives the value of mass flow rate of steam.

Throat and Exit Area

(a) If the nozzle is convergent then throat will be the exit

The throat area A_2 can be calculated as follows.

Let,

P_1 = Entrance pressure

P_2 = Exit or throat pressure

Enthalpy drop = $H_1 - H_2 kJ/kg$

Velocity at throat,

$$V_2 = 44.72\sqrt{(H_1 - H_2)} \ m/s$$

Then mass flow, $m = \frac{A_2 V_2}{x_2 V_{s_2}} kg/s$

Where,

A_2 = Throat area

V_{s2} = Specific volume of steam at pressure P_2 form steam table

x_2 = Dryness fraction of steam at pressure P_2

If the steam super-heated at throat, $m = \frac{A_2 V_2}{V_{sup}} kg/s$

Where,

$$V_{sup} = V_{s_2}\left(\frac{T_{sup_2}}{T_{s_2}}\right)$$

(b) If the nozzle is convergent - divergent

The exit area A_3 can be calculate by follows,

Enthalpy drop = $H_1 - H_3 kJ / kg$

Velocity at exit is V_3,

$$V_3 = 44.72\sqrt{\text{Enthalpy drop}}\ m/s$$

$$= 44.72\sqrt{(H_1 - H_3)}\ m/s$$

Then, mass flow m will be,

$$m = \frac{A_3 V_3}{x_3 V_{s_2}}$$

Where,

V_{s3} = Specific volume of steam at pressure P_3

X_3 = Dryness fraction of steam at pressure P_3.

Problems

1. Steam at a pressure of 10 bar and 0.98 dry is passed through a CD nozzle to a back pressure of 0.1 bar. The mass flow rate of steam is 0.55 kg/s. Let us determine,

- Pressure at throat.
- No. of nozzles used if each nozzle has a throat area of 0.5 cm^2.

The enthalpy drop used for reheating the steam by friction in the divergent part is 10% of overall isentropic drop. Take index of expansion n = 1.13.

Solution:

Given:

P_1 = 10 bar

x_1 = 0.98

P_2 = 0.1 bar

m = 0.55 kg/s

A_t = 0.52 cm^2

Since loss by friction is 10% the efficiency, η = 90%

n = 1.13

To Find:

- Throat pressure (P_t).
- No. of Nozzles (N).

Formula to be used:

$$m = \frac{A_t \times V_t}{v_t}$$

$$V_t = \sqrt{2000(h_1 - h_t)}$$

For saturated steam n = 1.135,

$$\frac{P_t}{P_1} = 0.577$$

P_t = 0.577 x 10

P_t =5.77 bar

Properties of steam:

At 10 bar, 9

h_1 = 2776.2 kJ/kg

S_1 = 6.583 kJ/kg-k

Now,

$$\text{No. of Nozzle required} = \frac{\text{Total mass flow rate}}{m}$$

$$m = \frac{A_t \times V_t}{v_t}$$

At 5.77 bar

$$v_t = 0.331205 \text{ m}^3 / \text{Kg}$$

$$h_t = 2753.25 \text{ KJ} / \text{Kg}$$

$$V_t = \sqrt{2000(h_1 - h_t)}$$

$$= \sqrt{2000(2776.2 - 2753.25)}$$

$$V_t = 214.24 \text{ m} / \text{s}$$

$$m = \frac{0.52 \times 10^{-4} \times 214.24}{0.331205}$$

$$m = 0.033\ Kg/s$$

$$N = \frac{0.55}{0.033}$$

N = 16.67 =17 Nozzles

Result

- P_t = 5.77 bar
- N = 17

2. A Steam enters a nozzle in a dry saturated condition and expands from a pressure of 2 bar to a pressure of 1 bar. It is observed that super-saturated flow is taking place and the steam flow reverts to a normal flow at 1 bar. Let us find the degree of under cooling, increase in entropy and the loss in the available heat drop due to irreversibility.

Solution:

Given:

P_1 = 2 bar

P_2 = 1 bar

Supersaturated flow

To find:

- Degree of under cooling.
- Increase in entropy.
- Loss in the available heat drop due to irreversibility.

Formula to be used:

$$\frac{T_2^1}{T_1} = \left(\frac{P_2}{P_1}\right)^{3/13},$$

Increase in entropy = $h_1 - h_3$,

Loss in the available heat drop due to reversibility $= \dfrac{h_1 - h_3}{\text{Degree of Undercooling}}$

From Mollier Chart,

For 2 bar and dry saturated:

h_t = 2700 kJ/kg

$h_3 = 2580$

$V_1 = 0.9\ m^3/kg$

$V_3 = 1.6\ m^3/kg$

$T_1 = 100°C.$

1. Degree of under cooling

From steam tables, for 2 bar, the initial temp is $T_1 = 120°C$.

T_2^1 = Temperature which super saturation occurs.

For supersaturated expansion,

$$\frac{P_1}{(T_1)^{13/3}} = \frac{P_2}{(T_2)^{13/3}} \text{ (or)}$$

$$\frac{T_2^1}{T_1} = \left(\frac{P_2}{P_1}\right)^{3/13}$$

$$\frac{T_2^1}{T_1} = \left[\frac{1}{2}\right]^{3/13} = 0.852$$

$T_2^1 = 102.24°C$

Degree of under cooling $= T_{sat} - T_2^1$

$= T_{20} - 102.24$

$= 17.76°C$

Increases in entropy $= h_1 - h_3$

$= 2700 - 2580$

$= 120$ kJ/ kg

Loss in the available heat drop due to reversibility $= \dfrac{h_1 - h_3}{\text{Degree of Undercooling}}$

$$= \frac{120}{17.76}$$

=6.756 KJ /Kg.

3. The flow rate through steam nozzle with isentropic flow from pressure of 13 bar was found to be 60 kg/min, steam is initially saturated. If the flow is super saturated, let us determine the throat area and increase in the flow rate.

Solution:

Given:

$P_1 = 13$ bar

$m = 60$ kg/min $= 1$ kg/s

Formula to be used:

$$s_2 = s_{f2} + x_2 S_{fg2}$$

$$V_2 = \sqrt{2000(h_1 - h_2)}$$

Throat area of Nozzle, $A_2 = \dfrac{m \times v_2}{V_2}$,

$$V_2 = \sqrt{\frac{2n}{n-1} \times p_1 v_1 \left[1 - \left(\frac{p_2}{p_1}\right)^{\frac{n-1}{n}}\right]}$$

Dry saturated steam,

Pressure at throat given by,

$P_2 = 0.577 \times P_1 = 0.577 \times 13$

$P_2 = 7.54$ bar

Properties of steam (from steam table) at 13 bar,

$h_1 = 2785.4$ kJ/kg

$s_1 = 6.491$ kJ/kgK

$v_1 = 0.15114$ m^3/kg

At 7.54 bar

$h_f = 710.26$ kJ/kg

$v_g = 0.254218$ m^3/kg

$s_{fg} = 4.658$ kJ/kgK

$h_g = 2054.81$ kJ/kg

$s_f = 2.0217$ kJ/kgK

1-2 isentropic expansion

$s_1 = 6.491$ kJ/kgK

$v_1 = 0.15114\ m^3/kg$

At 7.54 bar

$h_f = 710.26\ kJ/kg$

$v_g = 0.254218\ m^3/kg$

$s_{fg} = 4.658\ kJ/kgK$

$h_g = 2054.81\ kJ/kg$

$s_f = 2.0217\ kJ/kgK$

1-2 isentropic expansion,

$s_1 = s_2 = 6.491 KJ/Kg$

$S_2 = S_{f2} + x_2 S_{fg2}$

$6.491 = 2.0217 + x_2 4.658$

$x_2 = 0.9595$

$h_2 = 710.26 + 0.9595 \times 2054.81 = 2681.828 kJ/kg$

$v_2 = x_2 \times v_{g2}$

$= 0.9595 \times 0.254218 = 0.243922\ m^3/kg$

Velocity of steam at exit

$$V_2 = \sqrt{2000(h_1 - h_2)} = \sqrt{2000(2785.4 - 2681.828)}$$

$$V_2 = 455.13\ m/s$$

Throat area of Nozzle, $A_2 = \frac{m \times v_2}{V_2}$

$$= \frac{1 \times 0.243922}{455.13}$$

$$A_2 = 5.3594 \times 10^{-4} m^2$$

If the flow is super saturated $n = 1.3$

$$V_2 = \sqrt{\frac{2n}{n-1} \times p_1 v_1 \left[1 - \left(\frac{p_2}{p_1}\right)^{\frac{n-1}{n}}\right]}$$

$$= \sqrt{\frac{2\times1.3}{1.3-1}\times13\times10^5\times0.15114\left[1-\left(\frac{6.54}{13}\right)^{\frac{1.3-1}{1.3}}\right]}$$

$$= 448.49 \text{ m/s}$$

Specific volume at throat,

$$V_2 = V_1\left(\frac{p_1}{p_t}\right)^{1/n}$$

$$= 0.15114\left(\frac{14}{7.54}\right)^{\frac{1}{1.3}}$$

$$= 0.2298 \text{ m}^3/\text{kg}$$

Flow rate of steam through nozzle,

$$m = \frac{A_2 \times V_2}{V_2}$$

$$= \frac{5.34\times10^{-4}\times448.49}{0.2298}\times60$$

$$m = 62.88 \text{ Kg/min}$$

Increase in mass flow rate, m = 62.88 – 60 = 2.88 kg/min.

4

Steam Turbines, Condensers, Heat Transfer and Refrigeration System

4.1 Steam Turbines and Condensers

Turbines

A turbine is a rotary engine that extracts energy from a fluid flow. The simplest turbines have one moving part, a rotor-blade assembly. Moving fluid acts on the blades to spin them and impart energy to the rotor. Early turbine examples are windmills and waterwheels.

Classification based on mode of steam action:

- Impulse turbine.
- Reaction turbine.

Classification based on direction of steam flow:

- Axial flow turbine.
- Radial flow turbine.
- Mixed flow turbine.

Classification based on number of stages:

- Single stage turbine.
- Multistage turbine.

Classification based on the pressure of steam:

- Low pressure turbine.
- Medium pressure turbine.
- High pressure turbine.

Simple Impulse Turbine

The steam coming out of nozzle with a very high velocity impinges on the blades of the rotor in the turbine. The blades changes the direction of steam.

flow without changing the pressure. The examples for impulse turbines are De-Laval turbine,

Curtis turbine and Rateau turbine. Thus in an impulse turbine complete expansion of steam takes place in stationary nozzles and the velocity head i e kinetic energy is converted into shaft work or mechanical work in the turbine blades.

In short, in purely impulse turbine, the rotary motion of the shaft is obtained by having high velocity jets of steam directed against the blades attached to the rim of the rotor. Figure shows the schematic arrangement of nozzle and rotor in a simple impulse turbine with pressure velocity distribution.

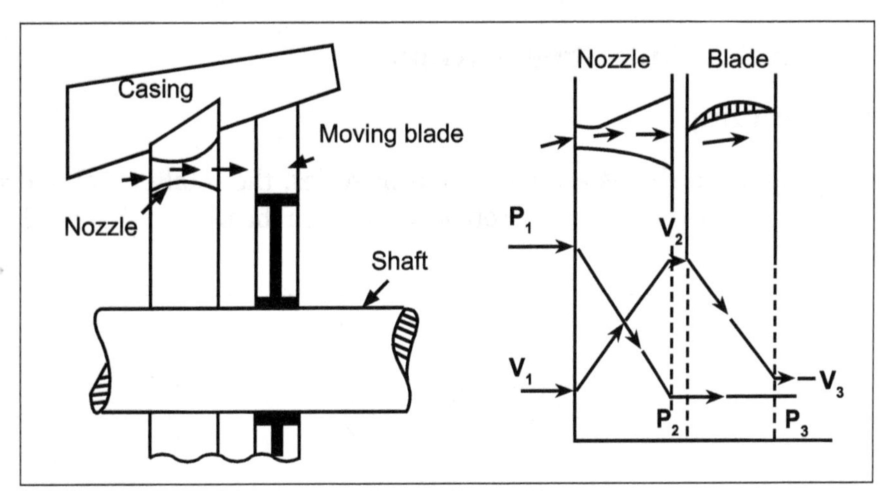

Simple Impulse Turbine.

In steam turbine terminology, an impulse turbine stage represents the combination of nozzle and moving blades. The nozzle is assumed to be in between states 1 and 2 and the blade is, in between states 2 and 3. The velocity of steam at nozzle inlet is low and can be considered as negligible.

Since the expansion through nozzle increases the kinetic energy, at nozzle exit velocity of steam is high i.e. $V_2 >> V_1$. Then the high velocity steam jet passes through the moving blades with no pressure drop but with gradual reduction in velocity. Hence $V_3 < V_2$. Increase in kinetic energy in the nozzle is obtained at the expense of pressure head. Hence $p_2 < p_1$. Also, expansion in the moving blade occurs with no change in pressure.

$$P_3 = P_2$$

Applications of Steam Turbine

- Fuel used are biomasses, coal etc.
- Steam heated processes in plants and factories.
- Modern steam turbines have automatic control system.
- Steam driven turbines in electric power plants.
- Because the turbine generates rotary motion, it is particularly suited to be used to drive an electrical generator about 90% of all electricity generation in the United States (1996).

4.1.1 Impulse Turbine, Pressure and Velocity Compounding, Velocity Diagram, Work Output, Losses and Efficiency

Compounding of steam turbines is the method in which energy from the steam is extracted in a number of stages rather than a single stage in a turbine. A compounded steam turbine has multiple stages i.e. it has more than one set of nozzles and rotors, in series, keyed to the shaft or fixed to the casing, so that either the steam pressure or the jet velocity is absorbed by the turbine in number of stages.

Velocity Diagrams for Multi-Stage Turbines

Pressure Compounding

When four simple impulse turbines are connected in series, the total enthalpy drop is divided equally among the stages. So, the pressure drop only occurs in the nozzle whereas there is no pressure drop in blades.

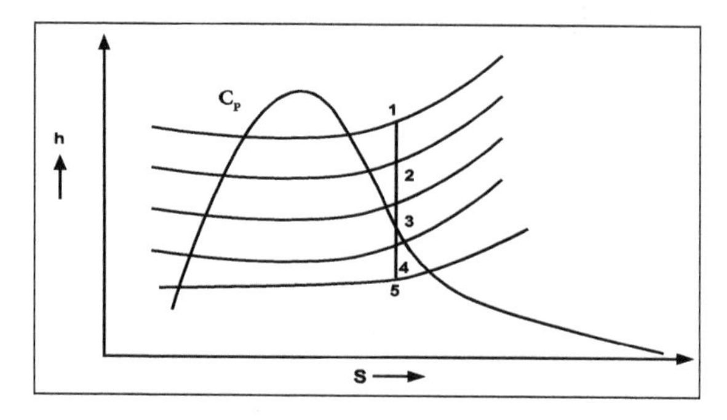

Enthalpy drop in each stage will be equal.

$$\therefore\ h_1 - h_2 = h_2 - h_3 = h_3 - h_4 = h_4 - h_5$$

So, $h_1 - h_2 = \dfrac{h_1 - h_5}{4}$

$\therefore$ the velocity of steam at exit from the first row of nozzle is given by,

$$V_1 = \sqrt{2000(h_1 - h_2)}$$

Where, h_1 and h_2 in kJ/kg

$$= \sqrt{2000\left(\frac{h_1 - h_5}{4}\right)}$$

$$= \frac{1}{2}\sqrt{2000(h_1 - h_5)} \qquad \text{...(i)}$$

But for a single stage turbine, the velocity of steam at exit of the nozzle,

$$V_1 = \sqrt{2000(h_1 - h_5)} \qquad \text{...(ii)}$$

From equation (i) and (ii) we can infer that the velocity of steam leaving the nozzles in each stage of 4-stage turbine is half of that for a single stage turbine. For 9-stage turbine, it will be one-third.

So, each impulse turbine operating at its maximum blading efficiency is given by,

$$\frac{V_b}{V_1} = \frac{\cos x}{2}$$

For n-stages, the enthalpy drop per stage will be:

$$(\Delta h)_{stage} = \frac{(\Delta h)_{total}}{n} = \frac{h_1 - h_n}{n}$$

Or

$$\text{Number of stages} = \frac{(\Delta h)_{total}}{(\Delta h)_{stage}}$$

Velocity Compounding

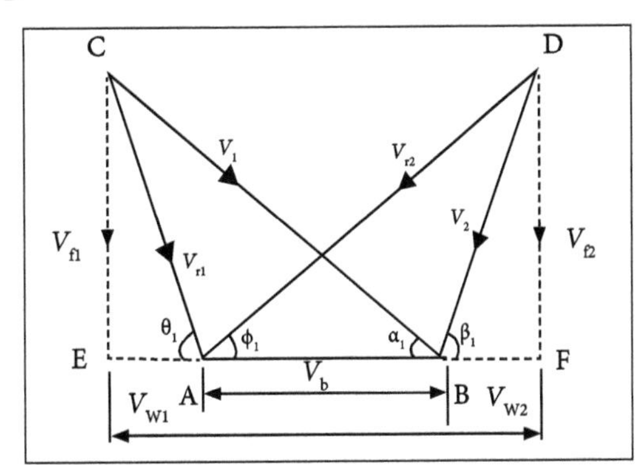

Velocity diagram for the first row of moving blades.

The kinetic energy of steam jets (1/2 m V_{12}) at nozzle exit is partially converted into work in the first row of moving blades with velocity differences from V_1 to V_2. Again kinetic energy $\left(\frac{1}{2}mV_3^2\right)$ of the exiting steam from the first row of moving blades is converted into work in the next row of moving blades and so on.

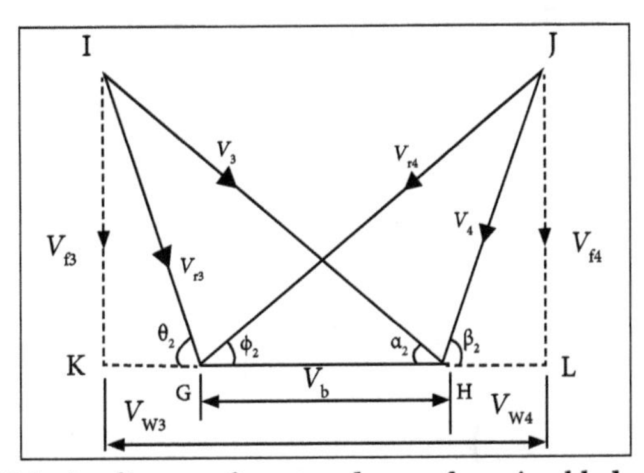

Velocity diagram for second row of moving blades.

Already we know that,

$$\frac{V_{r_2}}{V_{r_1}} = k$$

From this diagram,

$$\text{Work done} = m(V_{w1} + V_{w2})V_b$$

$$\text{Axial thrust}, F_{y1} = m(V_{f1} - V_{f2})$$

Kinetic energy of steam supplied for the first stage,

$$K.E_I = \frac{1}{2} m V_1^2$$

The same friction factor is considered for the next row of moving blades.

$$\therefore \frac{V_3}{V_2} = k \text{ And also } \frac{V_{r_4}}{V_{r_3}} = k$$

$$\text{Work done} = m\left(V_{w3} + V_{w4}\right)V_b$$

$$\text{Axial thrust, } F_{yII} = m\left(V_{f3} - V_{f4}\right)$$

Kinetic energy of stem supplied for second stage,

$$K.E_{II} = \frac{1}{2} m V_s^2$$

Total efficiency of steam turbine,

$$\eta = \frac{\text{Workdone}_I + \text{Workdone}_{II}}{K.E_I + K.E_{II}}$$

$$\therefore \eta = \frac{2V_b\left(V_{W_1} + V_{W_2} + V_{W_3} + V_{W_4}\right)}{V_1^2 + V_3^2}$$

Similarly, total axial thrust,

$$F_Y = F_{YI} + F_{YII}$$

$$= m\left(V_{f1} - V_{f2} + V_{f3} - V_{f4}\right)$$

$$= m\left[\left(V_{f1} + V_{f3}\right) - \left(V_{f2} + V_{f4}\right)\right]$$

Pressure and Velocity Compound

This method of compounding is the combination of two previously discussed methods. The total drop in steam pressure is divided into stage and the velocity obtained in each stage is also compounded. The rings of nozzle are fixed at the beginning of each stage and pressure remains constant during each stage. The change in pressure and velocity are shown in figure.

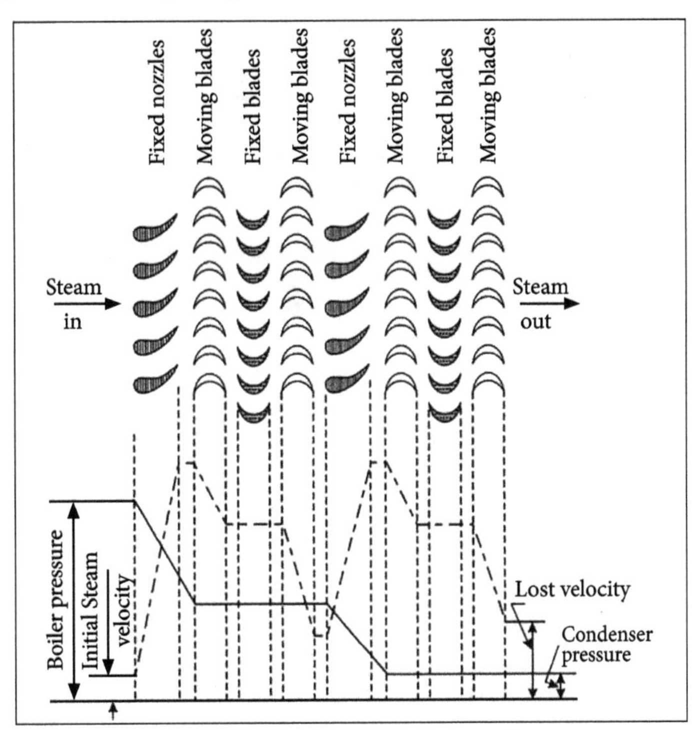

Pressure and velocity compound.

The energy losses in steam turbines are:

- Losses in regulating valves.
- Losses due to steam friction.
- Losses due to mechanical friction.
- Losses due to leakage.
- Losses due to wetness of steam.
- Losses due to radiation.
- Residual velocity losses.
- Carry over losses.

Problems

1. Steam issues from the nozzles of a De Laval turbine with a velocity of 1000 m/sec. The nozzle angle is 20°. Mean blade velocity is 400 m/sec. The blades are symmetrical. The mass flow rate is 1000 kg/h. Friction factor is 0.8, h_{nozzle} = 0.95. Let us determine,

- Blade angles.
- Axial thrust on the rotor turbine.
- Work done per kg of steam.
- Power developed.
- Blade efficiency.
- Stage efficiency.

Solution:

Given:

$V_1 = 1000$ m/s

$a = 20^o$

$V_o = 4000$ m/sec

$m = 1000$ kg/hs $= 0.27$ kg/sec

$\eta_{nozzle} = 0.95$

$$k = \frac{V_{r2}}{V_{r1}} = 0.8$$

To Find:

- Blade angle (θ, f).
- Axial thrust (F_y).
- Work done per kg of steam (W_D).
- Power developed (P).
- Stage efficiency (η_{stage}).
- Blade efficiency (η_{blade}).

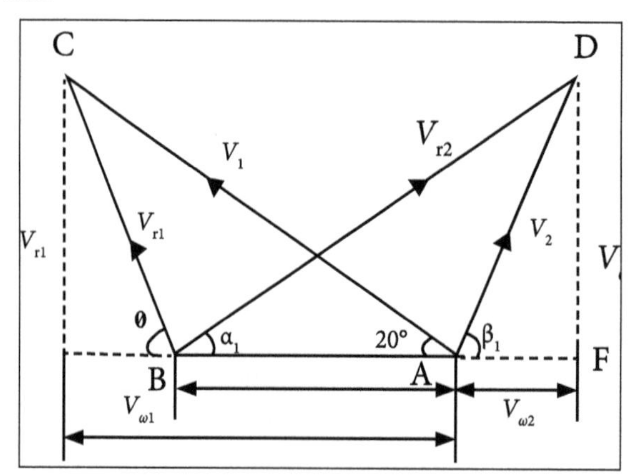

Formula to be used:

$$V_{\gamma 1} = \sqrt{V_{\gamma 1}^2 + (V_{\omega 1} - V_b)^2}$$

$$\tan\theta = \frac{V_{f1}}{V_{\omega 1} - V_b}$$

$$V_{f2} = V_{r2} \sin f$$

Driving force, $F_x = m(V_{\omega 1} + V_{\omega 2})$

Axial thrust, $F_y = m(V_{f1} - V_{f2})$

Power, $P = m\, V_b (V_{\omega 1} + V_{\omega 2})$

Work done/kg of steam, $W_D = V_b (W_{\omega 1} + V_{\omega 2})$

Blade efficiency, $\eta_b = \dfrac{m(V_{\omega 1} + V_{\omega 2}) \times V_b}{\dfrac{mV_1^2}{2}}$

Stage efficiency, $\eta_{stage} = \eta_b \times \eta_v$

From ΔBCE,

$$V_{\omega 1} = V_1 \cos 20° = 1000 \cos 20° = 939.69 \text{ m/s}$$

$$V_{f1} = V_1 \sin 20° = 1000 \sin 20° = 342.02 \text{ m/s}$$

From ΔACE,

$$V_{r1} = \sqrt{V_{f1}^2 + (V_{\omega 1} - V_b)^2} = \sqrt{342.02^2 + (939.69 - 400)^2}$$

$$V_{r1} = 638.93 \text{m/s}$$

$$\tan\theta = \frac{V_{f1}}{V_{\omega 1} - V_b} = \frac{342.02}{939.69 - 400}$$

$$\theta = 32.21$$

But,

$$\frac{V_{r2}}{V_{r1}}$$

$$V_{r2} = 0.8 \text{ x } 638.93$$
$$= 511.14 \text{ m/s}$$

For symmetrical blading,

$$\theta = f = 32^\circ \times 21$$

From ΔADF,

$$V_{f2} = V_{r2}\sin f$$

$$= 511.14 \times \sin 32^\circ.21'$$
$$= 273.50 \text{ m/s}$$

Similarly

$$V_b + V_{\omega 2} = V_{r2}\cos 32^\circ 21$$
$$400 + V_{\omega 2} = 511.14 \cos 32^\circ 21$$
$$V_{\omega 2} = 31.80 \text{ m/s}$$

From ΔBDF,

$$V_2 = \sqrt{V_{f2}^2 + V_{\omega 2}^2}$$

$$= \sqrt{273.50 + 31.80^2}$$

$$V_2 = 275.34 \text{ m/s}$$

Driving force,

$$F_x = m(V_{\omega 1} + V_{\omega 2})$$
$$= 0.27(939.69 + 31.80)$$
$$= 262.30 \text{ N}$$

Axial thrust,

$$F_y = m(V_{f1} - V_{f2})$$
$$= 0.27(342.02 - 273.50)$$
$$F_y = 18.5 \text{ N}$$

$$\text{Power, } P = m\, V_b (V_{\omega 1} + V_{\omega 2})$$
$$= 262.30 \times 400$$
$$= 104.92 \text{ kW}$$

Work done/kg of steam,

$$W_D = V_b(W_{\omega 1} + V_{\omega 2})$$
$$= 400\,(939.69 + 31.80)$$
$$W_D = 388.59 \text{ kJ}$$

Blade efficiency,

$$\eta_b = \frac{m(V_{\omega 1} + V_{\omega 2}) \times V_b}{\frac{mV_1^2}{2}}$$

Stage efficiency,

$$\eta_{stage} = \eta_b \times \eta_v$$
$$= 0.7771 \times 0.95$$
$$= 73.82\%$$

Result:

- $\theta = f = 32°21$
- $F_y = 18.5$ N
- $W_D = 388.59$ kJ
- $P = 104.92$ kW
- $\eta_{blade} = 77.71\%$
- $\eta_{stage} = 73.82\%$

2. A One stage of an impulse turbine consists of a converging nozzle ring and one ring of moving blades. The nozzles are inclined at 22° to the blades whose tip angles are both 35°. If the velocity of steam at exit from the nozzle is 660 m/s, let us calculate the blade speed so that the steam passes without shock. Let us also find the diagram efficiency neglecting losses if the blades are run at this speed.

Solution:

Given data:

$$\alpha = f = 22° \quad (\theta = \phi = q = b)$$
$$q = b = 35°$$
$$C_1 = V_a = 660 \text{ m/s}$$

To find:

- Blade speed.
- Diagram efficiency.

Formula to be used:

$$\eta_b = \frac{2V_D\left(V_{\omega 1} + V_{\omega 2}\right)}{V_1^2}$$

From the figure,

V_b = blade velocity (Scale: 1cm = 60 m/s)

= 4.2 × 60

= 252 m/s

Blade efficiency or diagram efficiency,

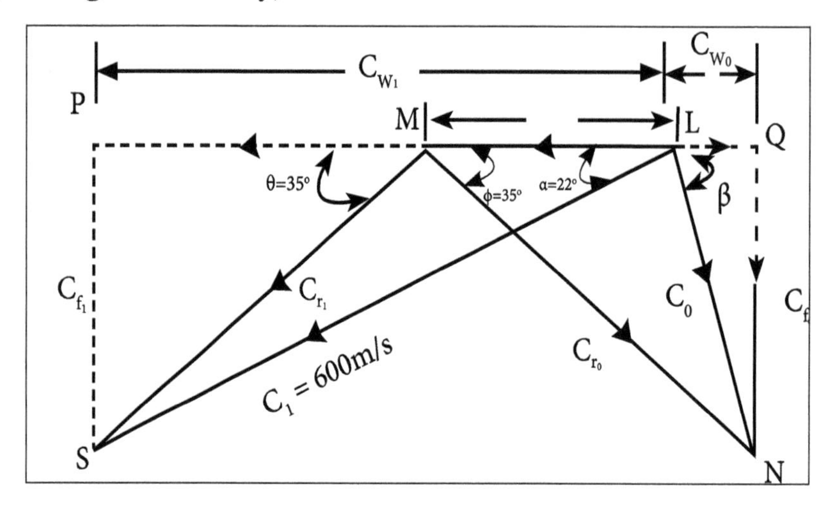

$$\eta_b = \frac{2V_D\left(V_{\omega 1} + V_{\omega 2}\right)}{V_1^2}$$

$$V_{\omega 1} = 10.2 \text{ x } 60 = 612 \text{ m/s}$$
$$V_{\omega 2} = 1.7 \text{ x } 60 = 102 \text{ m/s}$$

$$\eta_b = \frac{2 \times (612 + 102) \times 252}{660^2}$$

$$\eta_b = 82.6\%$$

3. A 300 kg/min of steam (2 bar, 0.98 dry) flows through a given stage of a reaction turbine. The exit angle of fixed blades as well as moving blades is 20° and 3.68 kW of power is developed. It the rotor speed is 360 rpm and tip leakage is 5 per cent; let us calculate the mean drum diameter and the blade height. The axial flow velocity is 0.8 times the blade velocity.

Solution:

Given data:

m_1 = 300 kg/min = 5 kg/sec

$N = 360$ rpm

$P = 2$ bar

$x = 0.98$

$V_f = 6.8\ V_b$

$P = 3.68$ kW $= 3.68 \times 10^3$ W

To find:

1) Drum diameter d,

2) Blade height, h.

Formula to be used:

$$V_b = \frac{\pi d_m N}{60}$$

$$BC = \frac{V_f}{\sin 20^\circ}$$

$$V_{\gamma_1} = \frac{V_{f1}}{\sin 20^\circ}$$

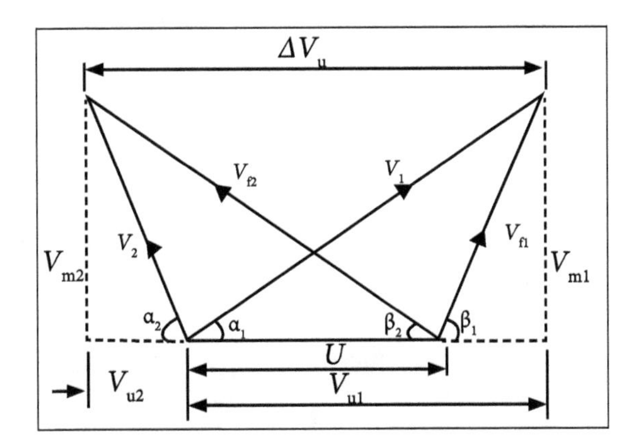

Power developed (P), $p = m(V_w + V_{w1})V_b$

Mass of steam flow, $m = \dfrac{\pi dm\ V_{f_1}}{x\ V_g}$

Drum diameter, $d = d_m - h$,

Since the tip leakage is 5 percent, therefore actual mass of steam flowing over the blades,

$$m = 5 - (5 \times 0.05)$$
$$= 4.75 \text{ kg/S}$$

We know that,

$$V_b = \frac{\pi d_m N}{60}$$

$$= \frac{\pi d_m \times 360}{60}$$

$$V_b = 18.849 \text{ dm m/s}$$
$$V_f = 0.8V_b$$
$$V_f = 0.8 \text{ x } 18.849 \text{ dm}$$
$$V_f = 15.0796 \text{ dm m/s}$$

We know that power developed (P), $p = m(V_w + V_{w1})V_b$.

$$3.68 \times 10^3 = 4.75 \times 57 \times 18.849 \text{ dm}$$
$$3.68 \times 10^3 = 5100.93 \text{ dm}^2$$
$$\text{dm}^2 = 1.3861$$
$$\text{dm} = 1.177 \text{ m.}$$
$$V_f = 15.0796 \text{ dm}$$
$$= 15.0796 \times 1.177$$
$$V_f = 17.753 \text{ m/s}$$

From steam tables, corresponding to a pressure of 2 bar, we find that specific volume of steam,

$$V_g = 0.885 \text{ m}^3 \text{/kg}$$

Mass of steam flow, $m = \dfrac{\pi dm\, V_{f_1}}{x\, V_g}$

$$4.75 = \frac{\pi \times 1.177 \times h \times 17.753}{0.98 \times 0.885}$$

$$h = 0.06257 \text{ m}$$

$$h = 627 \text{ mm}$$

Height of the blades, h = 62.7 mm

Drum diameter, $d = d_m - h$

$$= 1.177 - 0.0627$$

$$d = 1.114\text{m}$$

4. A Steam enters the blade row of an impulse turbine with a velocity of 600 m/s at an angle of 25° to the plane of rotation of the blades the mean blade speed is 250 m/s. The blade Angle at the exit side is 30°. The blade friction loss is 10%. Let us determine:

- The blade angle inlet.

- The work done per kg of steam.
- Blade efficiency.

Solution:

Given:

Steam velocity = 600 m/s,

Plane of rotation of the blades =25°,

Mean blade speed = 250 m/s,

Exit blade angle = 30°,

To Find:

- Blade angle inlet.
- The work done per kg of steam.
- Blade efficiency.

The velocity diagram is constructed as described below. AB is drawn to scale to represent the blade velocity. U=250 m/s.

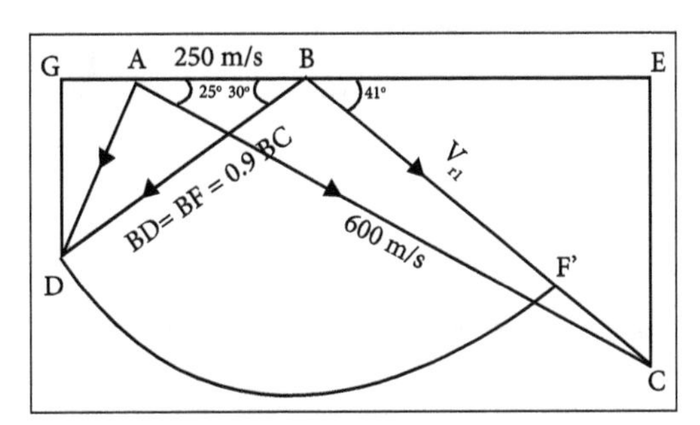

AC's drawn at 25° to AB to represent steam velocity. BC. Now represents the relative velocity V_{r1}. The angle CBE is measured as 41° and this is the blade exit angle.

(i) BF is marked as 0.9 BC to obtain the relative velocity at exit. BD is drawn at 30° to BA representing the blade angle at exit. Joining AD and drawing perpendicular to AB from D complete the velocity triangles.

GE is measured and converted to velocity. This length represents the change in the velocity of whirl.

This is found as 585 m/s.

(ii) The work done/kg $= U \times V_w = 250 \times 585$

$$= 146.25 \frac{kW}{kg}$$

(iii) Blade efficiency $= \frac{2UV_w}{V_{a_1}^2} = \frac{2 \times 250 \times 585}{600^2} = 81.25\%$

4.2 Impulse Reaction Turbine

The steam jets are directed at the turbine bucket shaped rotor blades where the pressure exerted by the jets causes the rotor to rotate and then the velocity of the steam to reduce as it imparts its kinetic energy to the blades.

The blades change the direction of flow of the steam, however its pressure remain constant as it passes through the rotor blades. The cross section of the chamber between the blades is constant. Impulse turbines are therefore also known as constant pressure turbines. The next series of fixed blades reverses the direction of the steam before it passes to the second row.

The rotor blades of the reaction turbine are delete arranged such that the cross section of the chambers formed between the fixed blades diminishes from the inlet side towards the exhaust side of the blades. The chambers between the rotor blades essentially form nozzles so that as the steam progresses through the chambers its velocity increases while at the same time its pressure decreases, just as in the nozzles formed by the fixed blades.

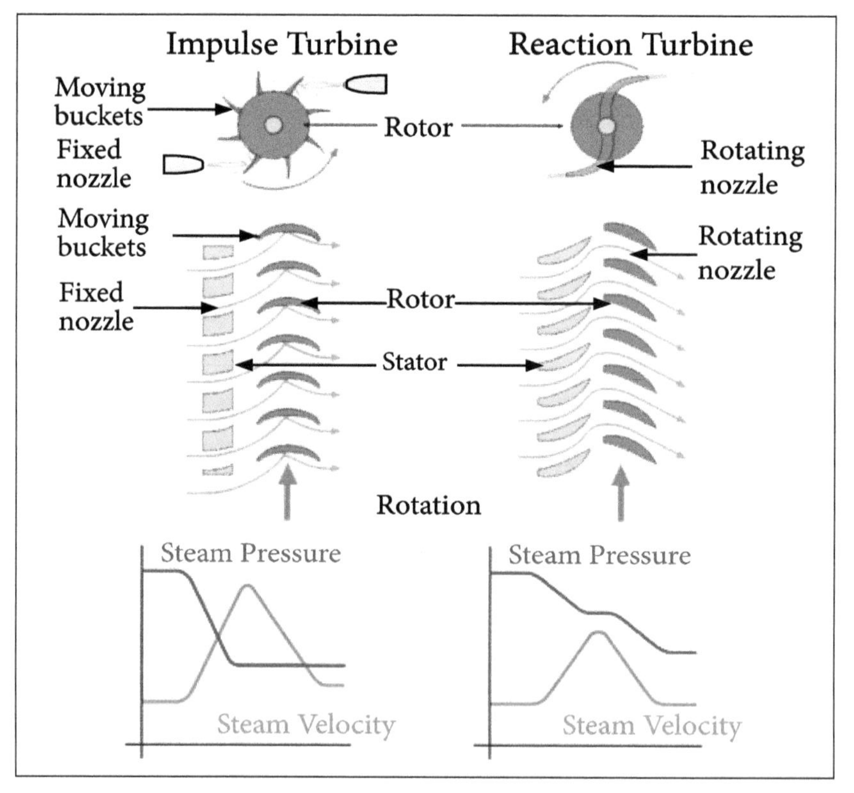

Impulse turbine. Reaction turbine.

Thus the pressure decreases in both the fixed and moving blades. The steam in a jet between the rotor blades creates a reaction force on the blades which in turn creates the turning moment on the turbine rotor, just as in steam engine.

Differences between Impulse and Reaction Turbines

S.No	Impulse Turbine	Reaction Turbine
1	In Impulse Turbine all hydraulic energy is converted into kinetic energy by a nozzle and it is is the jet so produced which strikes the runner blades.	In Reaction Turbine only some amount of the available energy is converted into kinetic energy before the fluid enters the runner.
2	The velocity of jet which changes, the pressure throughout remaining atmosphere.	Both pressure and velocity changes as fluid passes through a runner. Pressure at inlet is much higher than at outlet.
3	Water-tight casing is not necessary. Casing has no hydraulic function to perform. It only serves to prevent splashing and guide water to the tail race.	The runner must be enclosed within a water-tight casing.
4	The turbine is always installed above the tail race and there is no draft tube used.	Reaction turbine are generally connected to the tail race through a draft tube which is a gradually expanding passage. It may be installed below or above the tail race.
5	Impulse Turbine have more hydraulic efficiency.	Reaction Turbine have relatively less efficiency.
6	Impulse Turbine operates at high water heads.	Reaction turbine operates at low and medium heads.
7	Water flow in tangential direction to the turbine wheel.	Water flows in radial and axial direction to turbine wheel.
8	Needs low discharge of water.	Needs medium and high discharge of water.
9	Degree of reaction is zero.	Degree of reaction is more than zero and less than or equal to one.
10	Impulse turbine involves less maintenance work.	Reaction turbine involves more maintenance work.

4.2.1 Velocity Diagram, Degree of Reaction, Work Output, Losses And Efficiency

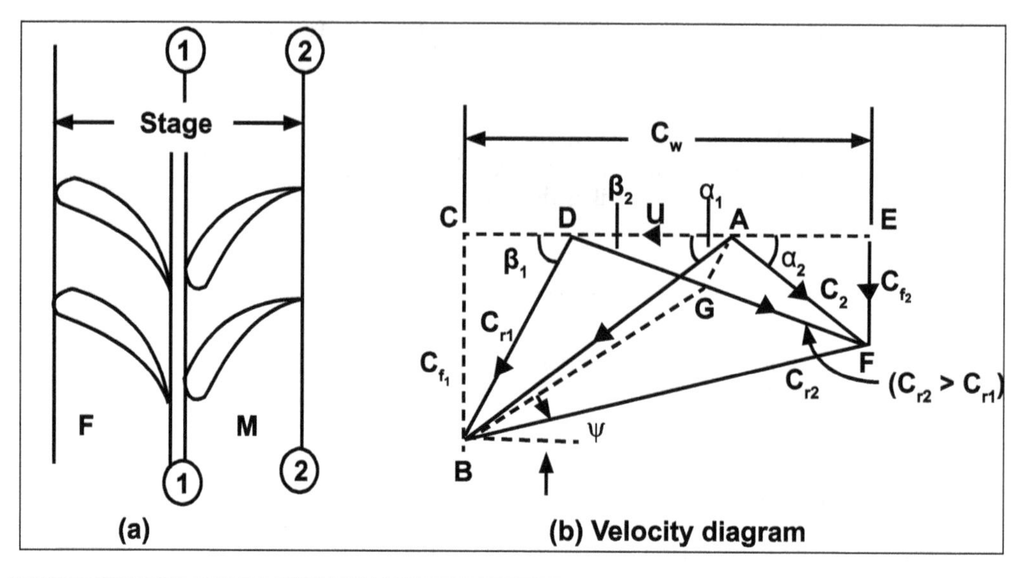

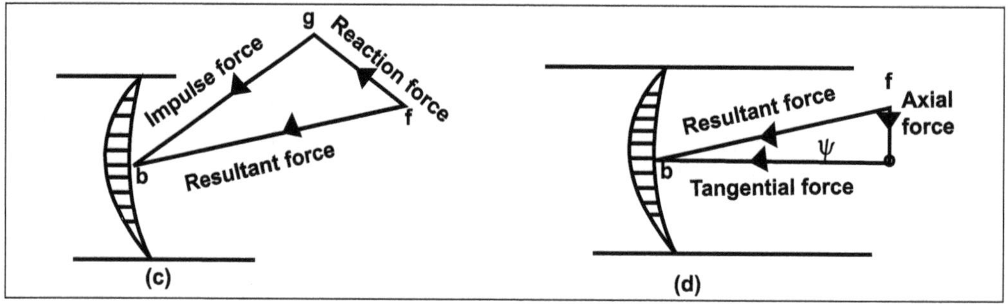

Degree of Reaction

The ratio of enthalpy drop in the moving blades to the total enthalpy drop in fixed and moving blades is known as the degree of reaction for reaction turbine stage as shown in the figure below.

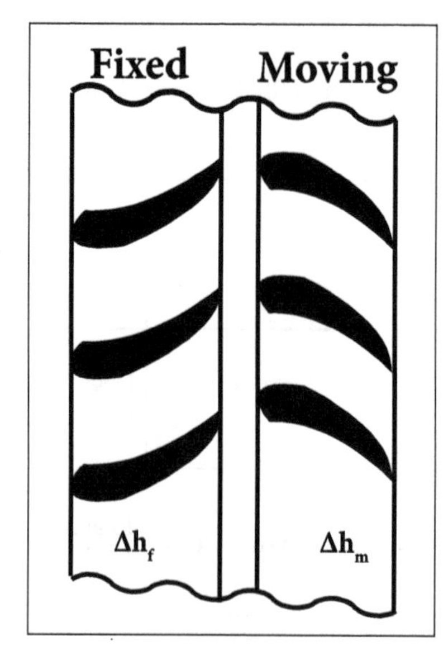

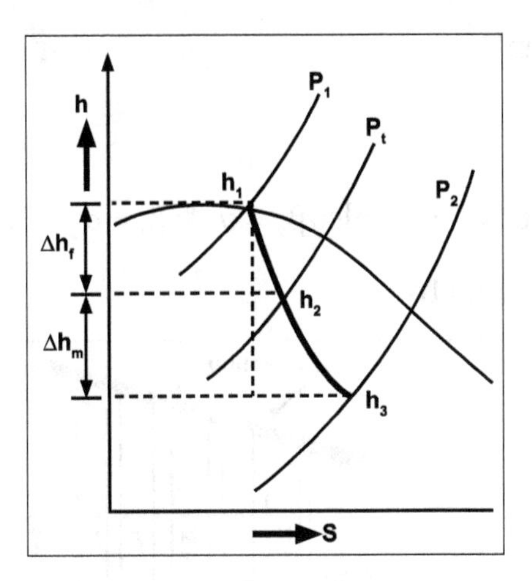

Thus,

$$R = \frac{\text{Enthalpy drop in moving blades}}{\text{total enthalpy drop in a stage}}$$

$$= \frac{\Delta h_m}{\Delta h_f + \Delta h_m}$$

Enthalpy drop in moving blades

$$\frac{V_{r2}^2 - V_{r1}^2}{2g_c}$$

Total enthalpy drop in a stage:

$$\Delta h_o = W = \Delta h_f + \Delta h_m = \frac{U}{g_c}(V_{u1} + V_{u2})$$

$$\therefore\ R = \frac{\left(V_{r2}^2 - V_{r1}^2\right)/2g_c}{U(V_{u1} + V_{u2})/g_c} = \frac{V_{r2}^2 - V_{r1}^2}{2U(V_{u1} + V_{u2})}$$

From velocity triangle: $V_{r2} = V_{f2}\ \mathrm{cosec}\,\beta_2$ & $V_{r1} = V_{f1}\ \mathrm{cosec}\,\beta_1$

And $(V_{u1} + V_{u2}) = V_{f1}\cot\beta_1 + V_{f2}\cot\beta_2$

$V_{f1} = V_{f2} = V_f$ (Because flow velocity is constant generally)

$$\therefore\ R = \frac{V_f^2}{2U}\frac{\mathrm{cosec}^2\beta_2 - \mathrm{cosec}^2\beta_1}{\cot\beta_1 + \beta_2} = \frac{V_f}{2U}\frac{\left(1+\cot^2\beta_2\right) - \left(1+\cot^2\beta_1\right)}{\cot\beta_1 + \cot\beta_2}$$

$$R = \frac{V_f}{2U}\frac{\cot^2\beta_2 - \cot^2\beta_1}{\cot\beta_1 + \cot\beta_2}$$

$$R = \frac{V_f}{2U}\left[\cot\beta_2 - \cot\beta_1\right]$$

If R= 0.5 i.e., 50% reaction, then

$$U = V_f\left[\cot\beta_2 - \cot\beta_1\right]$$

When R= 0.5, then $V_1 = V_{r2}$ and $V_2 = V_{r1}$ also $\beta_2 = \alpha_1$ & $\beta_1 = \alpha_2$,

Then the velocity Δ becomes symmetric.

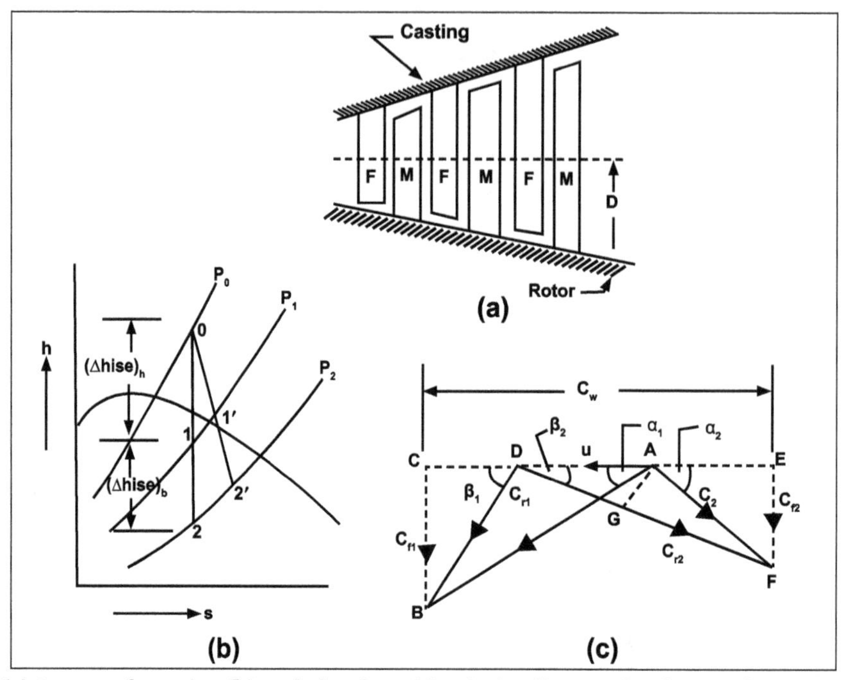

(a) Degree of reaction (b) enthalpy drop (c) velocity diagram for degree of reaction.

4.3 Jet and Surface Condensers

In jet condensers, the exhaust steam and cooling water come in direct contact and mix up together. Thus, the final temperature of condensate and cooling water leaving the condenser is same. The cooling water is sprayed on the exhaust steam to cause rapid condensation.

A jet condenser is very simple in design and cheaper than a surface condenser. It can be used when cooling water is cheaply and easily available. However, the condensate cannot be reused in the boiler, because it contains impurities like dust, oil, metal particles, etc.

Condenser may be Classified into the Following Two Groups

- Jet condenser.
- Surface condenser.

The jet condensers are also classified as:

- Low-level jet condenser.

- Counter-flow type.
- Parallel-flow type.
- High-level jet condenser.
- Ejector jet condenser.

1. Jet Condenser

The other name for this jet condenser is mixing type condenser. As its name implies the steam to be condensed and the cooling water come in direct contact and they mixed together. Therefore at the outlet of the condenser the temperature of cooling medium and the condensate are equal.

Since the two medium mixed together, the condensate cannot be used as feed water for the boiler. Because of this difficulty the jet condensers are not popular in today's world.

Depending upon the direction of flow of condensate the jet condensers are further divided into two types as:

- Parallel flow jet condenser.
- Counter flow jet condenser.

i) Parallel Flow Jet Condenser

In a parallel flow jet condenser, as its name implies, the condensate and cooling water flows in the same direction and are parallel to each other. In this type both condensate and cooling water enters from the top of the condenser and the mixture is removed from the bottom. A schematic representation of parallel flow jet condenser is shown in the figure. In this type exhaust steam from the turbine enters from the top of the condenser. Cooling water also enters the condenser from the top. The cooling water is made in the form of a spray with the help of a perforated tray.

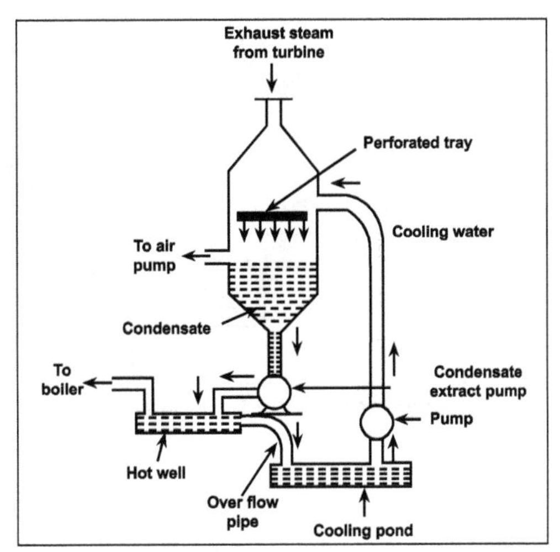

Parallel Flow Jet Condenser.

The cooling water condenses the exhaust steam and is collected as condensate in the bottom surface of the condenser. With the help of condensate extraction pump it is collected in the hot well and then it is made to feed as the feed water for the boiler in the next cycle of operation. In order to operate the condenser effectively an air pump is also used in the parallel flow jet condenser.

ii) Counter Flow Jet Condenser

In a counter flow jet condenser, the cooling water and the exhaust steam from the turbine enters in opposite direction. Here the cooling water enters from the top whereas the exhaust steam enters at the bottom of the condenser. In this case the air pump is connected at the top of the condenser which will maintain the vacuum within the condenser shell.

Cooling water is pumped by means of a pump to the perforated tray which is kept at the top of the shell. To increase the effectiveness of the condenser more than one perforated tray can be employed.

The exhaust steam from the turbine is fed at the bottom of the condenser shell. It is obvious that the temperature of exhaust steam is higher than the temperature of cooling water. Therefore the steam will tend to travel upwards.

Hence rapid condensation occurs when upward flowing exhaust steam and downward flowing cooling water are in contact. The condensate is collected at the bottom of the condenser shell which will be removed from there and collected in the hot-well with the help of a condensate extraction pump.

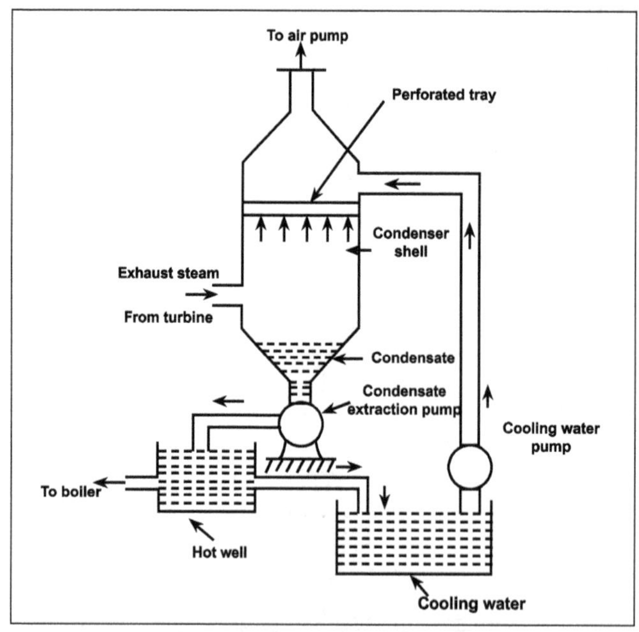

Counter Flow Jet Condenser.

Advantages of Jet Condenser

- The unit is simple and requires less capital cost.
- It needs less floor space.
- Requires less quantity of coolant i.e. cooling water.

Disadvantages of Jet Condenser

- Cooling water should be clean and free from impurities.
- High capacity air pump should be used.
- Piping is costly.
- Condensate cannot be reused.

2. Surface Condenser

In surface condensers, the exhaust steam and water do not come into direct contact. The steam passes over the outer surface of tubes through which a supply of cooling water is maintained. There may be single-pass or double-pass. In single-pass condensers, the water flows in one direction only through all the tubes, while in two-pass condenser the water flows in one direction through the tubes and returns through the remainder.

A jet condenser is simpler and cheaper than a surface condenser. It should be installed when the cooling water is cheaply and easily made suitable for boiler feed or when a cheap source of boiler and feed water is available. A surface condenser is most commonly used because the condensate obtained is not thrown as a waste but returned to the boiler.

The following is the main classification of surface condensers:

- Down-flow type.
- Central-flow type.
- Inverted-flow type.
- Regenerative type.
- Evaporative type.

(i) Down-Flow Type

In the figure is shown a down flow type of surface condenser. It consists of a shell which is generally of cylindrical shape, though other types are also used. It has cover plates at the ends and furnished with number of parallel brass tubes. A baffle plate partitions the water box into two sections.

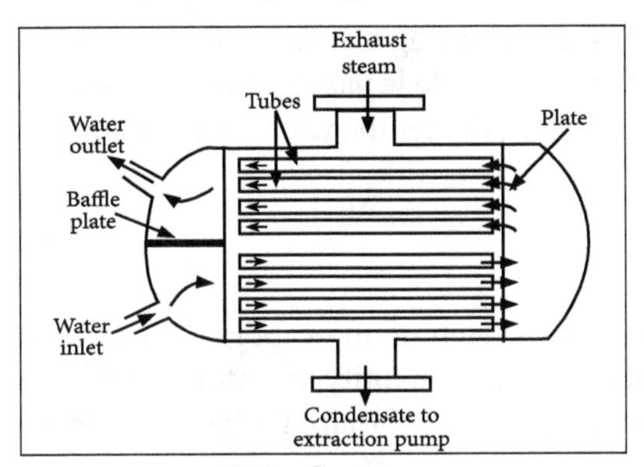

Down-flow type.

The cooling water enters the shell at the lower half section and after traveling through the upper half section comes out through the outlet. The exhaust steam entering shell from the top flows down over the tubes and gets condensed and is finally removed by an extraction pump. Due to the fact that steam flows in a direction right angle to the direction of flow of water, it is also called cross-surface condenser.

(ii) Central-Flow Type

In this type of condenser, the suction pipe of the air extraction pump is located in the centre of the tubes which results in radial flow of the steam. The better contact between the outer surface of the tubes and steam is ensured; due to large passages the pressure drop of steam is reduced.

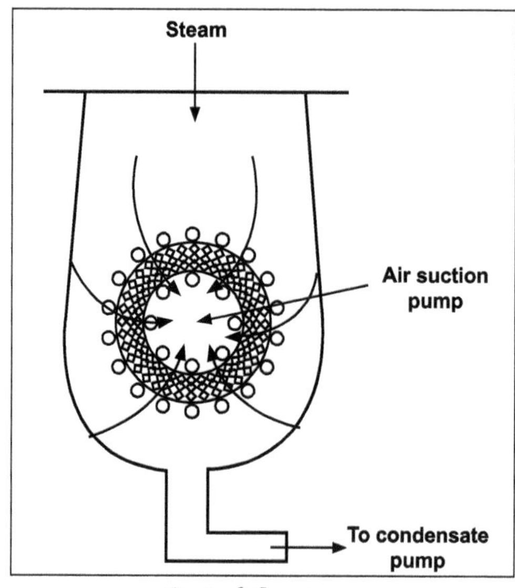

Central-flow type.

(iii) Inverted-Flow Type

This type of condenser has the air suction at the top, the steam after entering at the bottom rises up and then again flows down to the bottom of the condenser, by following a path near the outer surface of the condenser. The condensate extraction pump is at the bottom.

(iv) Regenerative Type

This type is applied to condensers adopting a regenerative method of heating of the condensate. After leaving the tube nest, the condensate is passed through the entering exhaust steam from the steam engine or turbine thus raising the temperature of the condensate, for use as feed water for the boiler.

(v) Evaporative Type

In the figure shows the schematic sketch of an evaporative condenser. The underlying principle of this condenser is that when a limited quantity of water is available, its quantity needed to condense the steam can be reduced by causing the circulating water to evaporate under a small partial pressure.

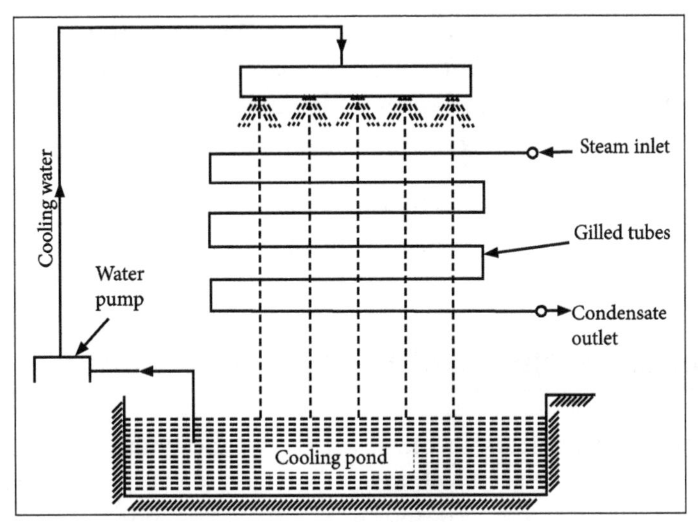

Evaporative type.

The exhaust steam enters at the top through gilled pipes. The water pump sprays water on the pipes and descending water condenses the steam. The water which is not evaporated falls into the open tank (cooling pond) under the condenser from which it can be drawn by circulating water pump and used over again.

4.3.1 Condenser Vacuum And Vacuum Efficiency

Condenser Efficiency

It is defined as the ratio of the difference between the outlet and inlet temperatures of cooling water to the difference between the temperature corresponding to the vacuum in the condenser and inlet temperature of cooling water, i.e.,

$$\text{Condenser efficiency} = \frac{\text{Rise in temperature of cooling water}}{\begin{bmatrix}\text{Temp. corresponding to} \\ \text{vacuum in the condenser}\end{bmatrix} - \begin{bmatrix}\text{Inlet temp. of} \\ \text{cooling water}\end{bmatrix}}$$

Or,

$$\text{Condenser efficiency} = \frac{\text{Rise in temperature of cooling water}}{\begin{bmatrix}\text{Temp. corresponding to the absolute} \\ \text{pressure in the condenser}\end{bmatrix} - \begin{bmatrix}\text{Inlet temp. of} \\ \text{cooling water}\end{bmatrix}}$$

Vacuum Efficiency

It is defined as the ratio of the actual vacuum to the maximum obtainable vacuum. The latter vacuum is obtained when there is only steam and no air is present in the condenser.

$$\text{Vacuum efficiency} = \frac{\text{Actual vacuum}}{\text{Maximum obtainable vacuum}}$$

$$= \frac{\text{Actual vacuum}}{\text{Barometer pressureo} - \text{Absolute pressure of steam}}$$

Note: In case of the absolute pressure of steam corresponding to the temperature of condensate being equal to the absolute pressure in the condenser, the efficiency would be 100%. Actually some quantity of air is also present in the condenser which may leak in and be accompanied by the entering steam.

The vacuum efficiency, therefore, depends on the amount of air removed by the air pump from the condenser.

4.4 Heat Transfer: Basic Modes of Heat Transfer

Heat is defined as energy in transition in the study of thermodynamics. It exists only when there is an exchange of energy between two systems or between a system and its surroundings. If the exchange of energy takes place due to a temperature difference, it is known as heat transfer.

Heat transfer is a universal phenomenon as temperature differences exist all over the universe. Its processes are involved in almost every engineering work.

It is essential to maintain thermal control of electronic components for optimum performance of electronic systems.

Heat transfer problems are also involved in the following fields:

- Air conditioning and refrigeration.
- Thermal and nuclear power plants.
- Internal combustion engines.
- Construction of dams.
- Structures and design and buildings.

There are three Modes of heat transfer:

- Conduction.
- Convection.
- Radiation.

Heat Transfer Occurs

It occurs due to temperature difference (high → low). The subject dealing with rate at which the heat flow Occurs is called heat transfer.

Importance: Design Equipment and Size Changes.

Example: Car Radiator.

Some example problems in Heat transfer are:

- Heat loss through thermal insulation on steam pipe.

- Heat transfer to water flowing through a tube.
- Heat transfer in an electric furnace.

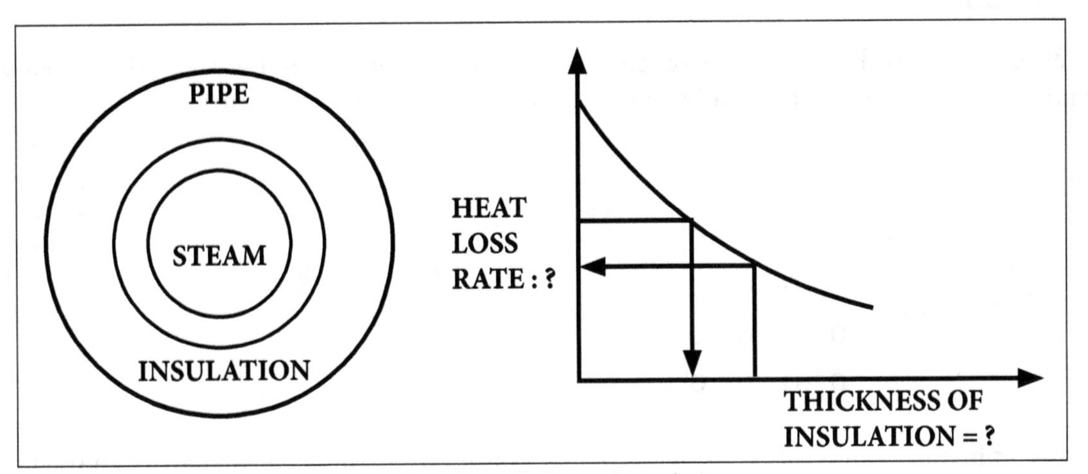

Heat loss through thermal Insulation.

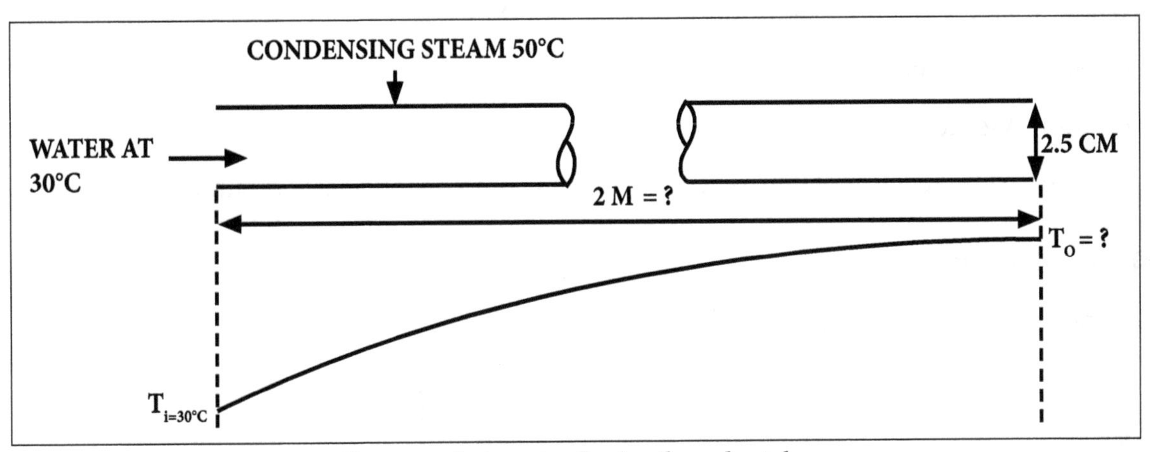

Heat transfer to water flowing through a tube.

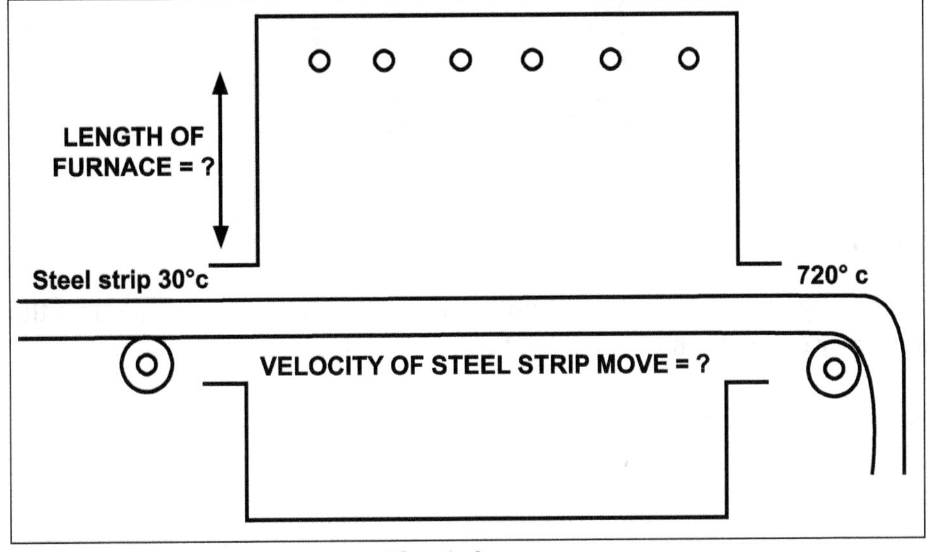

Electric furnace.

Conduction, Convection and Radiation

Conduction

Conduction is flow of heat in a substance due to exchange of energy between Molecules having more energy (Higher temp) and Molecules having less energy (Lower temp).

Solids

	O	O	O	O	O	
More temp	O	O	O	O	O	Less Temp
	O	O	O	O	O	
	O	O	O	O	O	

Molecules at higher temperature vibrate more compared to the molecules at lower temperature.

The below are the sequential steps for molecular vibration:

- Surface temperature is more.
- More Vibration.
- The first molecule will hit the second molecule and heat transfer occurs.

Liquid and Gases

These molecules will have random motion.

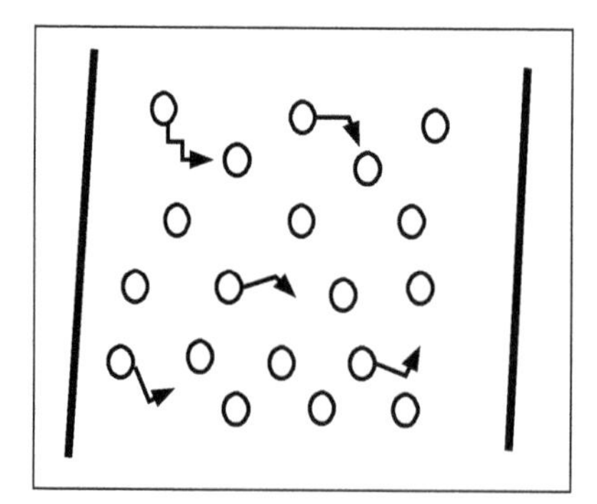

Convection

The transfer of energy from one region to another due to macroscopic motion in a fluid, added on to energy transfer by conduction is called heat transfer by Convection.

Two Types of Convection

Forced Convection

Fluid Motion Caused by an external agency.

Natural Convection

Fluid motion occurs due to density Variations caused by temperature differences.

Radiation

All physical matter emits thermal radiation in the form of electromagnetic waves because of vibrational and Rotational movements of the molecules and atoms which make up the matter.

Characteristics

- Rate of emission increases with temperature level.
- No material medium is required for energy transfer to occur.
- Mostly problems will come in all the three modes.

Example: Solar System.

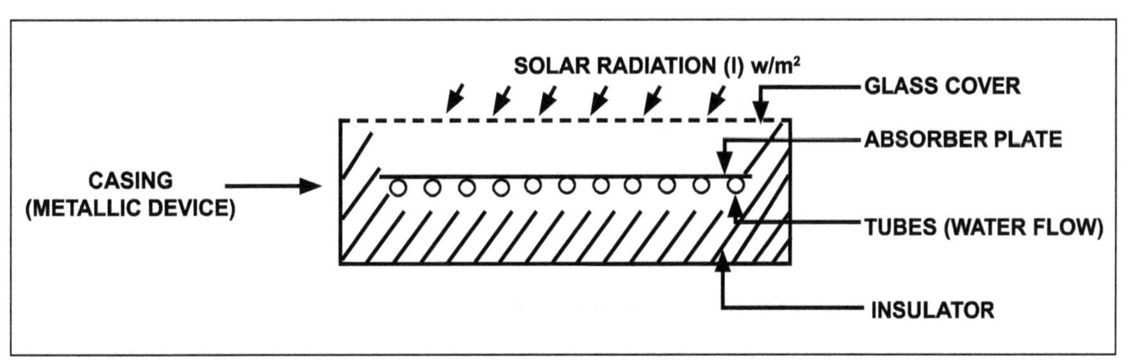

Solar Radiation Device.

$$I \times (\tau\alpha) \times A_p = q_u + q_{conduction} + q_{convection + radiation}$$

Where,

τ = Heat transmitted to glass Cover.

α = *Heat absorbed in plate.*

q_u = Heat gained (by convection).

$q_{conduction} + q_{convection + radiation}$ = Heat losses

4.4.1 One Dimensional Steady State, Conduction Through Slab, Cylinder and Sphere

In one-dimensional heat conduction i.e. temperature depends on one variable, we can devise a basic description of the process.

The first law in control volume form (steady flow energy equation) with no shaft work and no mass flow reduces the statement that $\sum Q$ for all surfaces=0 (no heat transfer on top or bottom of Figure) the heat transfer rate in the left at x is given by,

$$\dot{Q}(x) = -k\left(A\frac{dT}{dx}\right)_x \quad (1)$$

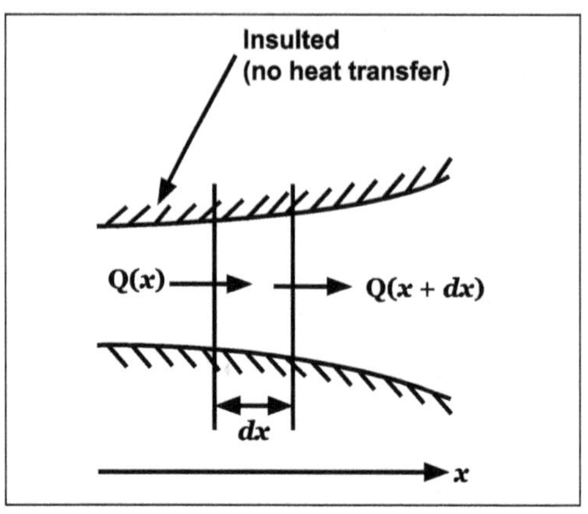

One-dimensional heat conduction.

The heat transfer rate on the right is given by.

$$\dot{Q}(x+dx) = \dot{Q}(x) + \left.\frac{d\dot{Q}}{dx}\right|_x dx + \ldots \quad (2)$$

Using the conditions on the overall heat flow and the expressions in (1) and (2),

$$\dot{Q}(x) - \left(\dot{Q}(x) + \frac{d\dot{Q}}{dx}(x)dx + \ldots\right) = 0 \quad (3)$$

Taking the limit as dx approaches zero we obtain the equation.

$$\frac{d\dot{Q}(x)}{dx} = 0 \quad (4)$$

Or,

$$\frac{d}{dx}\left(kA\frac{dT}{dx}\right) = 0 \quad (5)$$

If k is constant (i.e. if the properties of the bar are independent of temperature), this reduces to

$$\frac{d}{dx}\left(A\frac{dT}{dx}\right) = 0 \quad (6)$$

or (using the chain rule)

$$\frac{d^2T}{dx} + \left(\frac{1}{A}\frac{dA}{dx}\right)\frac{dT}{dx} = 0 \quad (7)$$

Equation (vi) or (vii) describes the temperature field for quasi-one-dimensional steady state (no time dependence) heat transfer.

Cylindrical Geometries

The approach illustrated for the analysis of the planar geometries also applies to cylindrical geometries provided that the change of the cross sectional area that is encountered by the radial heat flow is taken into account.

To illustrate this point, reference may be made to figure below.

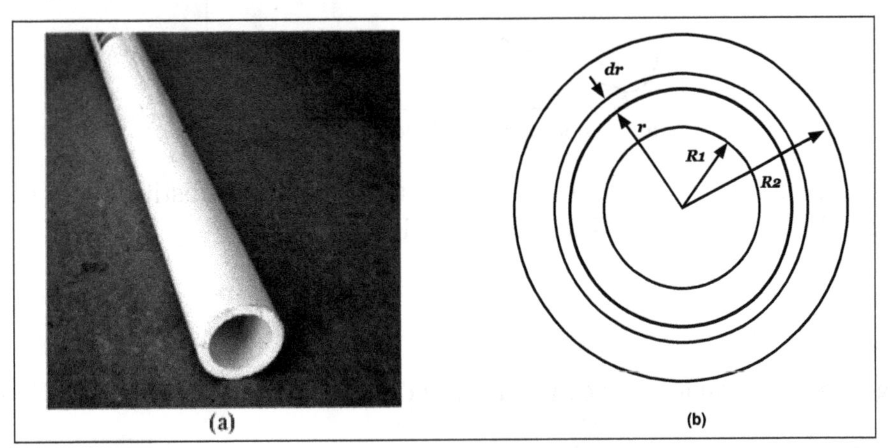

Hollow bore cylindrical tube. (a) Three-dimensional, (b) Cross-sectional view for analysis.

The tube has an inner radius R_1 and an outer radius R_2. This nomenclature, shown in the (b) part of the figure, is enhanced by the control volume outlined in red. The inner and outer radii of the control volume are, respectively r and $(r+dr)$.

The temperature it the surface of the how of the tube be T_{hi} and the temperature at the outside surface of the tube T_{io}, These temperatures acute a radians outward heat now Q In the steads state Q is independent of the radial position.

If Q were to vary with r, the temperature would necessarily vary with time. At any moment of time, let Q(r) the magnitude of the heat flow that is entering the control volume at the radius r. By the same token let $Q(r + dr)$ represent the heat flow that is leaving the control volume at a radius $(r+dr)$ in the steady state.

$$Q(r) = Q(r+dr) \qquad ...(i)$$

Or,

$$\frac{dQ}{dr} = 0 \qquad ...(ii)$$

Next, the radial heat flow is expressible by mean of Fourier's Law as:

$$Q = -kA(r)\frac{dT}{dr} \qquad ...(iii)$$

The area A(r) is illustrated in figure. The magnitude of A(r) is equal to $2\pi rL$

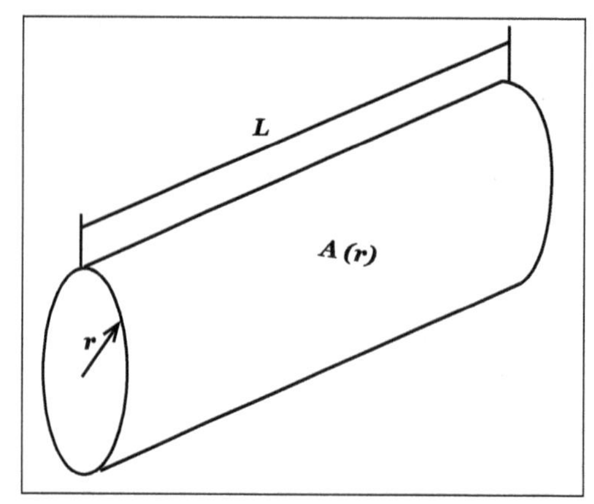

The area A(r) normal to the direction of radial heat flow.

Introduction of the expression for A(r) into Equation (iii) and the substitution of the resulting for Q into the energy-balance equation, Equation (ii), yields:

$$\frac{d}{dr}\left[-k \times 2\pi rL\frac{dT}{dr}\right]=0 \qquad \text{...(iv)}$$

If k can be regarded as a constant or can be replaced by an average value k as in Equation (iv) becomes:

$$\frac{d}{dr}\left[r\frac{dT}{dr}\right]=0 \qquad \text{...(v)}$$

By inspection,

$$r\frac{dT}{dr}=C_1 \qquad \text{...(vi)}$$

And,

$$dT=\frac{C_1 r}{r} \qquad \text{...(vii)}$$

The integration of this equation yields,

$$T(r) = C_1 In(r)+C_2 \qquad \text{...(viii)}$$

The integration constant C_1 and C_2 are determined by applying the boundary condition that $T(r=R_1)=T_{hi}\,(r=R_2)=T_{lo}$. The end result of this step is:

$$\frac{T(r)-T_{lo}}{T_{hi}-T_{lo}}=\frac{In(r/R_2)}{In(R_1/R_2)} \qquad \text{...(ix)}$$

The temperature distribution represented by Equation (ix) is plotted in figure. It can be seen from the figure that the distribution departs from the straight line that represents the variation in a planar one-dimensional geometry.

The slope of the distribution for the cylindrical case is greatest near the inside radius and decreases as the outer radius is approached.

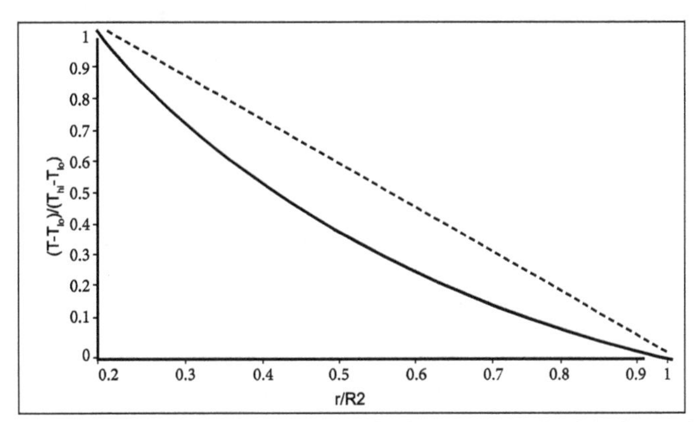

Temperature distribution across the thickness of a cylindrical annulus.

The rate of heat transfer Q passing through the cylindrical annulus can be obtained by making use of Equation (iii) and the temperature distribution from Equation (ix).

When the indicated operations are performed then,

$$Q = 2\pi Lk \frac{(T_{hi} \quad T_{lo})}{\text{In}(R_2 \quad R_1)}$$

From this, it follows that the thermal resistance of a cylindrical annulus is:

$$R = \frac{\text{In}(R_2 - R_1)}{2\pi Lk}$$

Conduction Through Slab

Heat Conduction in Solids for Closed System (Or) Control Volume by taking the fundamental laws:

- Law of Conservation of mass trivially satisfied.
- Newtons 2nd law of motion trivially satisfied.

So we need to Satisfy only first law of thermodynamics,

$$\frac{d\theta}{dt} - \frac{dw}{dt} = \frac{dE}{dt}$$

(Usually No Work is Involved)

$$\frac{d\theta}{dt} = \frac{dE}{dt}$$

To get familiar we take one dimensional Steady State Situations:

It means the temperature does not vary with time. The temp Varies only in one direction.

(i) Infinite Slab

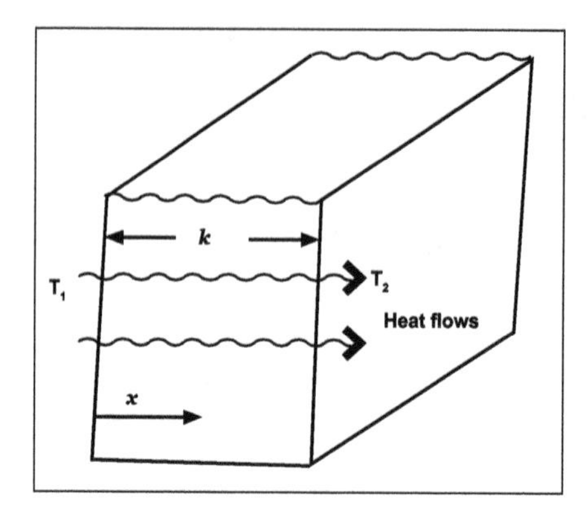

We need to find:

- The rate at which the head flows across this Solid.
- And the temperature distribution in solid.

Taking one dimensional,

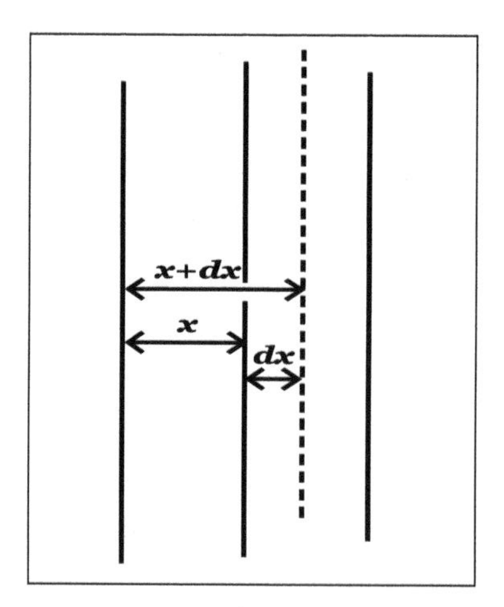

$$T = f(x)$$

$$\frac{q}{A} = -k\frac{dT}{dx}$$

So, $\frac{q}{A}$ = Constant (Using I[st] law of Thermodynamics)

By Integrating from 0 to x,

$$\frac{q}{A}dx = -k\frac{dT}{dx}$$

$$\frac{q}{A}\int_{0}^{b} dx = -k\int_{T_1}^{T_2} \frac{dT}{dx}$$

$$\frac{q}{A}(b) = -k(T_2 - T_1)$$

$$\frac{q}{A} = \frac{k}{b}(T_2 - T_1)$$

$$\frac{T_1 - T}{T_1 - T_2} = \frac{x}{b}$$

(ii) Infinitely Long Hollow Cylinder

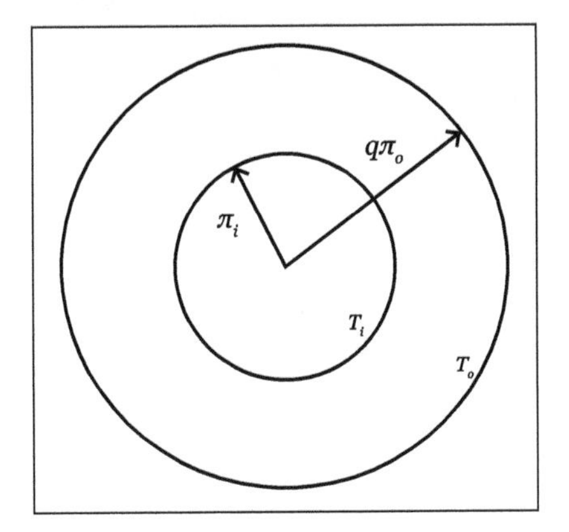

K = Constant

To Find:

- Temperature distribution.
- Radial heat flow rate.

For one dimensional flow

$$T=f(r)$$

$$\frac{q}{A} = -k\frac{dT}{dr}$$

$$\frac{q}{2\pi rL} = -k\frac{dT}{dr}$$

$$\frac{q}{2\pi rL}\int_{r_i}^{r_o}\frac{dr}{r} = -k\int_{T_i}^{T_o} dT$$

So the Corresponding results,

$$q = \frac{2\pi Lk(T_i - T_o)}{\ln(r_o / r_i)}$$

$$\frac{T-T_i}{T_o-T_i}=\frac{\ln(r/r_i)}{\ln(r_o/r_i)}$$

A long hollow to be with:

ID = 2 cmT_i=70°c

OD = 4 cmT_o=100°c

Heat flow Rate per Unit Length:

$$\frac{q}{QL}=?$$

$$k = 0.58\ w/mk$$

Formula:

$$\frac{q}{L}=\frac{2\pi k(T_i-T_o)}{\ln(r_o/r_i)}$$

$$=\frac{2\times\pi\times0.58(70-100)}{\ln\frac{2}{1}}$$

$$\frac{q}{L}=-157.7\,w/m$$

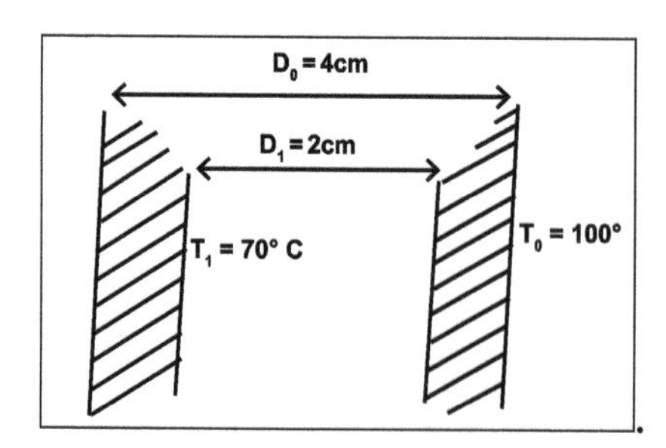

Heat Flow Through Sphere

Consider a hollow sphere of inner radius r_1 and outer radius r_2. Let T_1 be the temperature at the inner surface and T_2 be the temperature at the outer surface. Assume that $T_1 > T_2$, so that heat will flow from inside to outside.

Consider a spherical element at any radius r (between r_1 and r_2) of thickness dr. Then rate of heat flow according to Fourier's law is given as,

$$Q=-k(4\pi r^2)\frac{dT}{dr}\ (i)$$

Where,

$A = 4\pi r^2$ -Area of heat transfer.

k = Thermal conductivity of a material of which sphere is made.

Rearranging equation (i), we get:

$$\frac{dr}{r^2} = \frac{-4\pi k}{Q} dT \qquad \text{...(ii)}$$

Integrate equation (ii) between limits:

When,

$r = r_1$

$T = T_1$

$r = r_2$

$T = T_2$

$$\int_{r_1}^{r_2} \frac{dr}{r^2} = \frac{-4\pi k}{Q} \int_{T_1}^{T_2} dT$$

$$\left[-\frac{1}{r}\right]_{r_1}^{r_2} = \frac{-4\pi k}{Q}(T_1 - T_2)$$

$$\left[\frac{1}{r_1} - \frac{1}{r_2}\right] = \frac{-4\pi k}{Q}(T_1 - T_2)$$

On Rearranging, we get:

$$Q = \frac{4\pi k(T_1 - T_2)}{\left[\frac{1}{r_1} - \frac{1}{r_2}\right]}$$

$$Q = \frac{4\pi r_1 r_2 k(T_1 - T_2)}{(r_2 - r_1)} \qquad \text{...(iii)}$$

$rm = \sqrt{r_1 r_2}$ = Mean radius which geometric mean for sphere.

Equation (iii) we get,

$$Q = \frac{4\pi r_m^2 k(T_1 - T_2)}{(r_2 - r_1)}$$

Problems

1. Let us determine the heat transfer through the plane of length 6m, height 4m and thickness 0.30 m and also about the temperature of inner and outer surface are 100°C and 40°C.The thermal conductivity of wall is 0.55 W / m k.

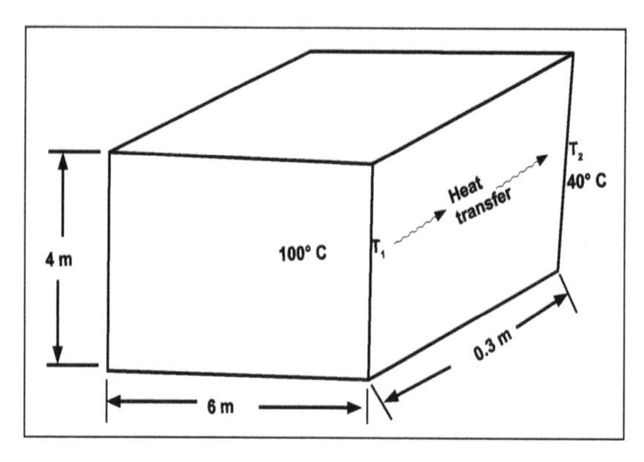

Solution:

Given:

Inner Surface Temperature T_1 = 100°C + 273 = 373 k

Outer Surface temperature T_2 = 40°C + 273 = 313 k

Thickness, b = 0.30 m

Area, A = 6 × 4 = 24 m^2

Thermal Conductivity, K = 0.55 W / m k

To Find:

Heat transfer (q)

We know that, heat transfer through plane wall is

$$q = \frac{KA}{b}(T_1 - T_2)$$

$$= \frac{0.55 \times 24}{0.30}(373 - 313)$$

$$= 2640 \text{ w}.$$

Result: Heat transfer Q = 2640 W.

2. Let us consider a wall of 0.6 m thickness having thermal Conductivity of 1.2 W / m k and it is to be insulated with a material having an average thermal Conductivity of 0.3 W / m k. Inner and Outer Surface temperature are 1000°C and 10°C respectively. It heat transfer rate is 1400 W / m^2. Also calculate the thickness of Insulation.

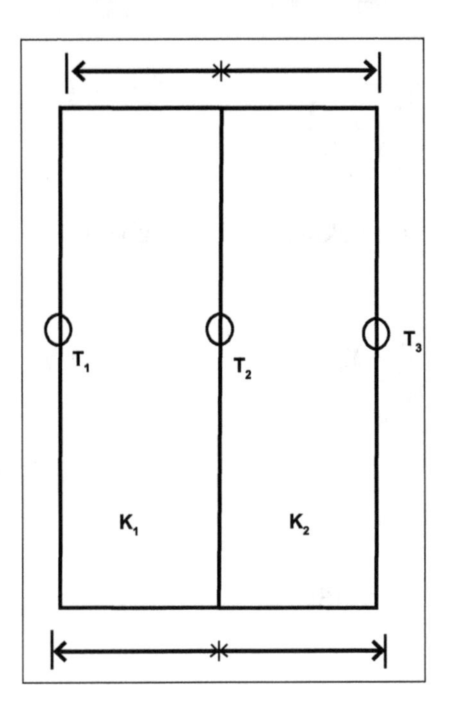

Solution:

Given:

Thickness of wall, b_1 = 0.6 m

Thermal Conductivity of wall K_1 = 1.2 W / m k

Thermal Conductivity of Insulation K_2 = 0.3 W / m k

Inner Surface temperature, T_1 = 1000C + 273 = 1273 k

Outer Surface temperature, T_3 = 10°C + 273 = 283 k

Heat transfer per Unit Area $\frac{q}{A}$ = 1400 W / m²

To Find:

Thickness of Insulation (b_2)

Formula to be used:

$$\Rightarrow \frac{q}{A} = \frac{T_1 - T_3}{\frac{b_1}{k_1} + \frac{b_2}{k_2}}$$

We know that,

$$q = \frac{T_1 - T_3}{R_{th}} \left[\therefore R_{th} = \frac{1}{h_1 A} + \frac{b_1}{k_1 A} + \frac{b_2}{k_2 A} + \frac{b_3}{k_3 A} + \frac{1}{h_2 A} \right]$$

$$q = \frac{T_1 - T_3}{\frac{1}{h_1 A} + \frac{h_1}{k_1 A} + \frac{b_2}{k_2 A} + \frac{b_3}{k_3 A} + \frac{1}{h_2 A}}$$

Heat transfer Co efficient h_1, h_2 and thickness b_3 is not given.

So, neglect that terms then we get:

$$q = \frac{T_1 - T_3}{\frac{b_1}{k_1 A} + \frac{b_2}{k_2 A}} \Rightarrow \frac{q}{A} = \frac{T_1 - T_3}{\frac{b_1}{k_1} + \frac{b_2}{k_2}}$$

By substituting the values,

$$1400 = \frac{1.273 - 283}{\frac{0.6}{1.2} + \frac{b_2}{0.3}} \Rightarrow b_2 = 0.0621\,\text{m}$$

Result:

Thickness of Insulation, L_2 = 0.0621 m

3. Let us consider a Composite wall of 10 cm thick layer of building brick, K = 0.7 W / m k and 3 cm thick plaster, k = 0.5 W / m k. and an insulating material of k = 0.08 W / m k is to be added to reduce the heat transfer through the wall by 40%. Let us calculate the thickness of insulation.

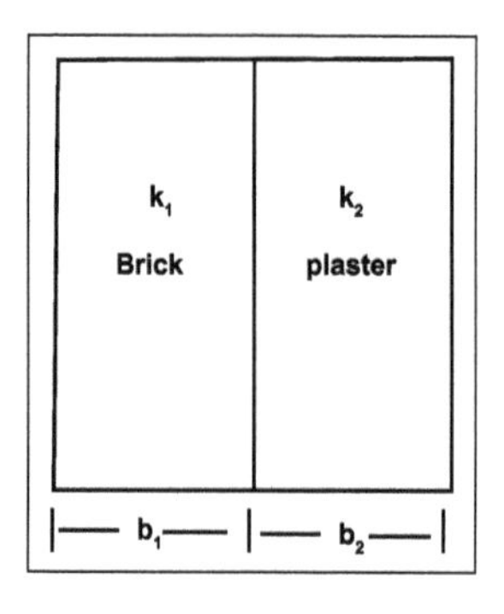

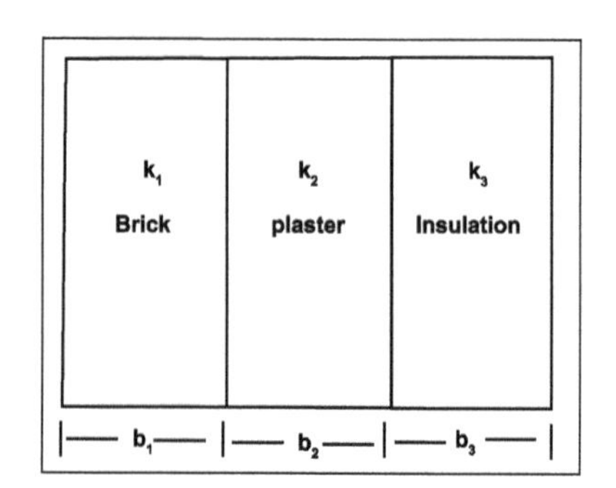

Solution:

Given:

Thickness of brick b_1 = 10 cm = 0.1 m

Thermal Conductivity of brick k_1 = 0.7 W / m k

Thickness of plaster b_2 = 3 cm = 0.03 m

Thermal Conductivity of plaster k_2 = 0.5 W / m k

Thermal Conductivity of Insulation k_3 = 0.08 W / m k

To Find:

Thickness of Insulation to reduce the heat loss through the wall by 40% (L_3)

Formula to be used:

$$q=\frac{T_1-T_2}{\frac{b_1}{K_1}+\frac{b_2}{K_2}}$$

$$\text{Heat flow rate}, Q=\frac{T_1-T2}{R_{th}}$$

Where,

$$R_{th}=\frac{1}{A}\left[\frac{1}{h_1}+\frac{b_1}{K_1}+\frac{b_2}{2}+\frac{b_3}{K_3}+\frac{1}{f_{12}}\right]$$

The term h_1 and h_2 are not given. So neglect that terms,

$$R=\frac{1}{A}\left[\frac{b_1}{K_1}+\frac{b_2}{K_1}+\frac{b_3}{K_3}\right]$$

Considering two Slabs (i.e.) neglect b_3 term:

$$q=\frac{T_1-T_2}{\frac{b_1}{K_1}+\frac{b_2}{K_2}}$$

Taking $\left[\because A=1m^2\right]$

$$=\frac{(115-75)-(70-50)}{\ln\left[\frac{115-75}{70-50}\right]}$$

[Assume heat transfer q=100w]

$$T_1-T_2=20.28\text{ k}$$

Heat loss to be reduced by 40% due to Insulation;

So, heat transfer is 60 W,

$$q=\frac{T_1-T_2}{R_{th}}$$

$$q = \frac{T_1 - T_2}{\frac{1}{A}\left[\frac{b_1}{k_1} + \frac{b_2}{k_2} + \frac{b_3}{k_3}\right]}$$

$$60 = \frac{20.28}{\frac{1}{1}\left[\frac{0.1}{0.7} + \frac{0.03}{0.5} + \frac{b_3}{0.08}\right]}$$

$$\left[A = 1\ m^2\right]$$

$$60\left[\frac{0.1}{0.7} + \frac{0.03}{0.5} + \frac{b_3}{0.08}\right] = 20.28$$

$$\frac{0.1}{0.7} + \frac{0.03}{0.5} + \frac{b_3}{0.08} = 0.338$$

$$\frac{b_3}{0.08} = 0.135 \Rightarrow b_3 = 0.0108$$

Result:

Thickness of Insulation, $b_3 = 0.0108$ m

4.5 Basic Theory of Radiant Heat Transfer, Black Body, Monochromatic Radiation and Total Emissive Power

Planck's Distribution Law

The relationship between the monochromatic emissive power of a black body and wave length of a radiation at a particular temperature is given by the following expression:

$$E_{b\lambda} = \frac{c_1 \lambda^{-5}}{\ell\left[\frac{c_2}{\lambda T}\right]_{-1}}$$

Where,

$E_{b\lambda}$ = Monochromatic emissive power (W/m²).

λ = Wavelength (M).

$C_1 = 0.374 \times 10^{-15}\left(W - m^2\right)$

$C_2 = 14.4 \times 10^{-3}\left(m\ k\right)$

Wien's Displacement Law

The Wien's law gives the relationship between temperature and wavelength corresponding to the Maximum Spectral emissive Power of the black body at the temperature.

$$\lambda_{Max}T = 2898\ \mu\ mk \qquad \left[\mu = 10^{-6}m\right]$$

$$\lambda_{Max}{}^{xT} = 2.9 \times 10^{-3}mk$$

Stefan – Boltzmann Law

The emissive Power of a black body is proportional to the fourth power of absolute temperature.

$$E_b \alpha\ T^4$$

$$E_b = \sigma\ T^4$$

Where,

E_b = Emissive Power $\left(W/m^2\right)$

σ = Stefan – Boltzmann Constant

$\sigma = 5.67 \times 10^{-8}$ (W/m² k)

T = Temperature (k)

Radiation

The heat transfer from one body to another without any transmitting medium is known as Radiation. It is an electromagnetic waves phenomenon. All types of electromagnetic waves are classified in terms of wavelength and propagated at the speed of light, i.e. 3×10^8 m/s.

E_b = Black body radiation or emissive power (W/m²)

σ = Stefan Boltzmann Constant ($5.67\ 10^{-8} W/m^2K^4$)

T = Temperature in K

$$E_b = \sigma T^4$$

Concept of Black Body

Black body is an ideal surface having the following properties: A black body absorbs all incident radiation, regardless of wavelength and direction. For a prescribed temperature and wave length, no surface can emit more energy than black body.

A black body is regarded as a perfect absorber of Incident radiation. A black body Condition can be approached in practice by forming a cavity in a material as shown in the above diagram. Radiation passing through the hole into the cavity is repeatedly absorbed and reflected at the cavity walls until it all absorbed.

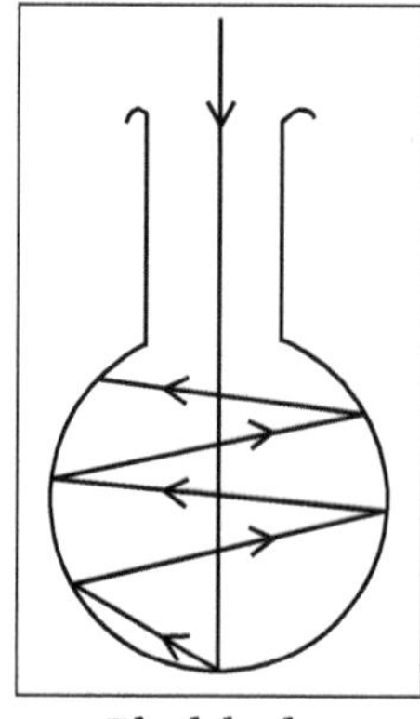

Block body.

A black body is a perfect emitter. This is a fact which can be proved as follows. Consider a black body at a uniform temperature, placed inside an arbitrarily shaped. Perfectly insulated enclosure composed of another black body whose temperature is also uniform but different from that of the former. The black body and the enclosure will reach a common equilibrium temperature after a period of time due to heat transfer.

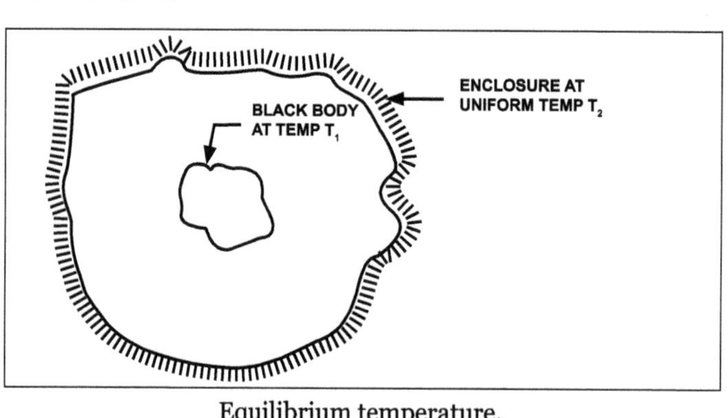

Equilibrium temperature.

The heat transfer from one body to another without any transmitting medium is known as radiation. It is an electromagnetic waves phenomenon. All types of electromagnetic waves are classified in terms of wavelength and propagated at the speed of light, i.e. 3×10^8 m/s.

Emissivity

The emissivity of a surface is the ratio of the radiation emitted by the surface at a given temperature to the radiation emitted by a blackbody at the same temperature. The radiation emitted by actual surfaces is less than that emitted by the blackbody.

The value of ε is in the range $0 \leq \varepsilon \leq 1$, is a measure of how closely a surface approximates a blackbody. The emissivity of real surfaces varies with the temperature of the surface, the wavelength and the direction of the emitted radiation. Surfaces with high emissivity are also very absorptive, that is, they will readily absorb radiation striking them.

These properties may vary depending on the wavelength of radiation falling on the surface. For the example, the surface may reflect much of the visible radiation falling on it, but not much of the ultraviolet (UV) radiation or infrared radiation falling on it.

It considers radiative heat exchange between two small, gray bodies, 1 and 2. The term 'small', denotes mean that their size is very small compared to the distance between them. Let the emissivities of surfaces 1 and 2 be ε_1 and ε_2 respectively and their absorptivities be α_1 and α_2 respectively.

Since the surfaces are gray (not black), surely we have to consider the effect of multiple reflections. However, implication of 'small' body is that the portion of radiation emitted by either body that is reflected by the other body is considered to be 'lost' in space and does not return to the originating surface.

Energy emitted by body 1 and incident on body 2 = $F_{12}.A_1.\varepsilon_1.T_1^4$

Of this energy amount absorbed by body 2 = $\alpha_2.F_{12}.A_1.\varepsilon_1.\sigma.T_1^4$

Therefore, energy transferred from body 1 to body 2;

(Since $\alpha_2 = \varepsilon_2$),

$$Q_1 = \varepsilon_1.\varepsilon_1.A_1.F_{12}.\sigma.T_1^4$$

Similarly energy transferred from body 2 and 1 is;

$$Q_2 = \varepsilon_1.\varepsilon_1.A_2.F_{21}.\sigma.T_2^4$$

Net radiant energy exchange between 1 and 2 is given by,

$$Q_{12} = \varepsilon_1.\varepsilon_{2.}A_1.F_{21}.\sigma.\left(T_2^4 - T_2^4\right) = \varepsilon_{1.}\varepsilon_{2.}A_2\sigma.\left(T_1^4 - T_1^4\right)$$

Since $A_1.F_{12} = A_2.F_{21}$

The product ($\varepsilon_{1.}\varepsilon_2$) is known as 'equivalent emissivity (ε_{eq})' for a system of two 'small' gray bodies.

Emissive Power (E_b)

The emissive power is defined as the total amount of radiation emitted by a body per unit time and unit area. It is expressed in W/m².

$$E_b = \sigma T^4 \left(W/m^2\right)$$

Where,

$$\sigma = 5.67 \text{ x}10^{-8} W/m^2$$

Monochromatic Massive Power (Eb_2)

The energy emitted by the surface at a given length per unit time per unit area in all directions is known as monochromatic emissive power.

Emissivity

It is defined as the ability of the surface of a body to radiate heat. It is also defined as the ration of the emissive power of anybody to the emissive power of a black body of equal temperature.

$$\alpha = \frac{E}{E_b}$$

4.5.1 Heat Exchangers

Types of Heat Exchanger

There are several types of heat exchangers which may be classified on the basis are:

- Nature of Heat Exchange Process.
- Relative Direction of Fluid Motion.
- Design and Constructional Features.
- Physical State of Fluids.

(i) Nature of Heat Exchange Process

On the basis of the nature of heat exchange process, heat exchangers are classified as,

- Direct contact heat exchangers; e.g.:- Cooling tower.
- Indirect contact heat exchangers; e.g.:- Air Pre heater, Economics.

(ii) Relative Direction of Fluid Motion

(a) Parallel Flow Heat Exchanger

This heat exchanger, the hot and the cold fluids enter at the same end of the heat exchanger and flow through in the same direction and leave together at the other end as shown in Figure.

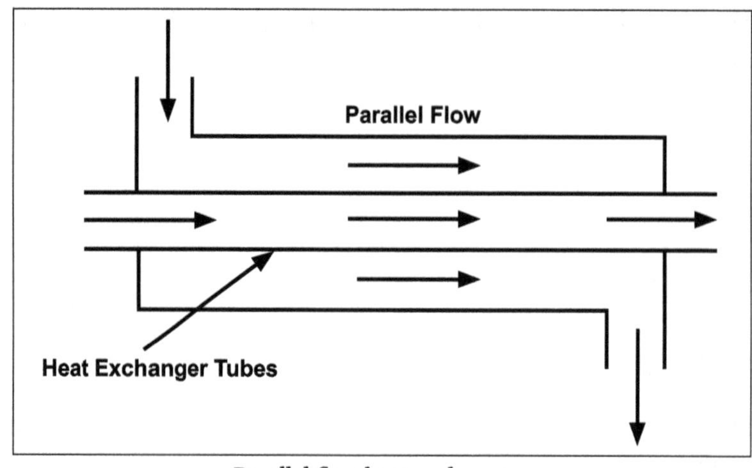

Parallel flow heat exchanger.

(b) Counter flow heat exchanger

In this heat exchanger hot and cold fluids enter in the opposite ends of the heat exchanger and flow in opposite directions as shown in Figure below.

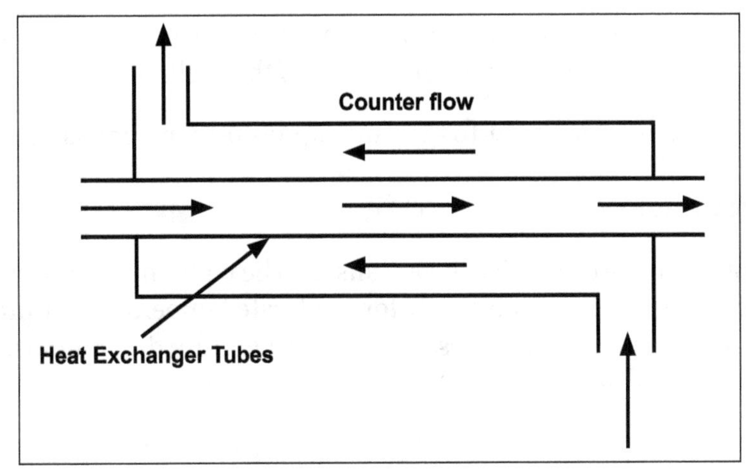

Counter flow heat exchanger.

(c) Cross Flow Heat Exchanger

In this heat exchanger, the two fluids flow at right angles to each other as shown in Figure. In this arrangement the flow may be mixed or unmixed.

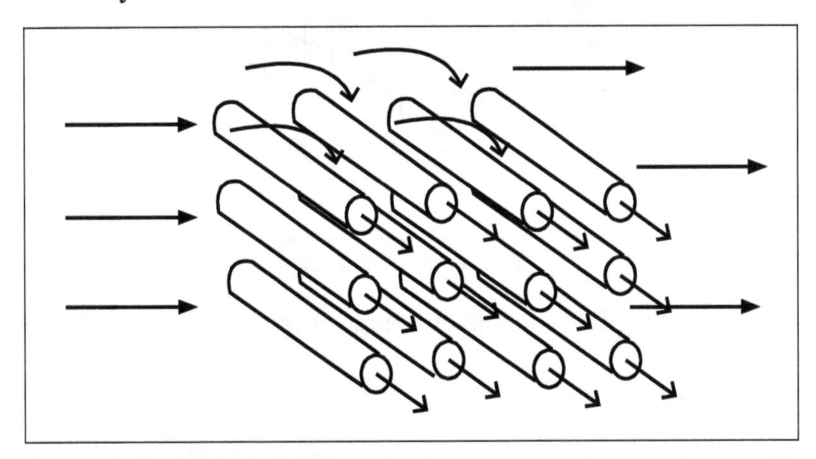

(iii) Design and Constructional Feature

On the basis of design and constructional features, the heat exchangers are classified as follows.

- Tubular heat exchangers.
- Plate heat exchangers.
- Plate fin heat exchangers.
- Tube-fin heat exchangers.
- Regenerative heat exchangers.

a) Tubular Heat Exchangers

The tubular heat exchangers are available in many sizes, flow arrangements and types. They can withstand a wide range of operating pressures and temperatures.

A commonly used design is shell-and-tube heat exchanger which consists of round tubes mounted on the cylindrical shells with their axes parallel to that of the shell.

The combination of fluids may be liquid-to -gas and liquid-to-liquid or gas-to-gas.

b) Plate heat exchangers

In these types thin plates are used to affect heat transfer. The plates may be either smooth or corrugated. These heat exchangers are suitable only for moderate temperature or pressure as the plate geometry restricts the use of maximum pressure and temperature differentials.

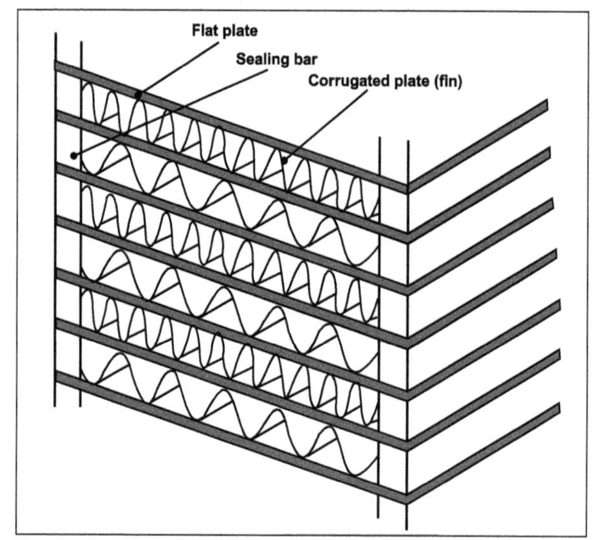

Plate heat exchangers.

The compactness factor for the plate exchangers ranges from 120 to 230 m^2/m^3.

c) Plate Fin Heat Exchangers

This heat exchangers use louvered or corrugated fins separated by flat plates. It can be arranged on both side of the plate to get cross-flow, parallel-flow or counter-flow arrangements.

These heat exchangers are used for gas-to-gas applications at low pressures and temperatures not exceeding 800°C.

They also find use in cryogenic applications. The compactness factor for these heat exchangers is up to 6000 m^2/m^3.

d) Tube-Fin Heat Exchangers

A heat exchanges are used when a maximum operating pressure or an extended surface is needed on one side. The tubes may be either round or flat.

Tube-fin heat exchangers are used in gas-252 Heat and Mass Transfer turbine, nuclear, fuel cell, automobile, heat pump, refrigeration, airplane, Cryogenics etc.

The operating pressure is about 30 atm. And the operating temperature ranges from minimum cryogenic temperatures to about 870 Dc. The maximum compactness ratio is about 330 m^2/m^3.

e) Regenerative Heat Exchangers

The regenerative heat exchangers may be either static type or dynamic type. The static type has no moving parts and consists of a porous mass like balls, powders, pebbles, etc. through which hot and cold fluids pass alternatively.

Example: Air preheaters used in coke manufacturing and glass melting plants. In dynamic type regenerators, the matrix is arranged in the form of a drum which rotates about an axis in such a manner that a given portion of the matrix passes periodically through the hot stream and then through the cold stream.

The heat absorbed by the matrix from the hot stream is transferred to the cold stream during its run.

(iv) Physical State of Fluids

Based on the physical state of fluids inside the exchanger, heat exchangers are classified as:

- Condensers.
- Evaporators.

(a) Condensers

The condensing fluid remains at constant temperature throughout the exchanger. The temperature of the colder fluid is gradually increased from inlet to outlet.

(b) Evaporators

The cold fluid remains at constant temperature. The temperature of hot fluid gradually decreases from inlet to outlet.

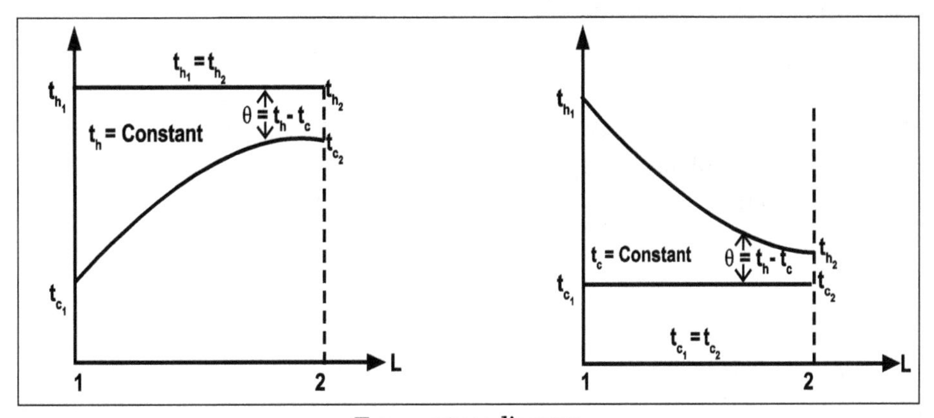

Temperature diagram.

Problems

1. Assume that sun to be a black body is emitting radiation with maximum intensity at $\lambda = 0.5\ \mu$. Let us calculate its surface temperature and emissive power.

Solution:

Given:

Wavelength $\lambda = 0.5\ \mu$

$$\lambda = 0.5\ \mu \times 10^{-6}\ m[1\mu = 10^{-6}\ m]$$

To Find:

- Surface temperature.
- Emissive Power.

Formula to be used:

Emissive Power, $E_b = \sigma T^4$

(a) According to Wien's displacement law,

$$\lambda_{max} T = 2.9 \times 10^{-3} m\ k$$

$$0.5 \times 10^{-6} \times T = 2.9 \times 10^{-3}$$

$$T = 5800\ k \rightarrow \text{Surface temperature}$$

(b) According to Stefan – Boltzmann law,

Emissive Power, $E_b = \sigma T^4$

$$E_b = 5.67 \times 10^{-8} \times (5800)^4$$
$$E_b = 64.1 \times 10^6 W/m^2$$

Result:

- Surface temperature T = 5800 k.
- Emissive Power $E_b = 64.1 \times 10^6 W/m^2$.

2. The problem based on the given statement that the sun emits Maximum Radiation at $\lambda = 0.52\ \mu$. Assuming the sum to be a black body, let us calculate the surface temperature of the sun. Also calculate the Monochromatic emissive Power of the Sun's Surface.

Solution:

Given:

$$\lambda_{max} = 0.52\ \mu$$
$$= 0.52 \times 10^{-6} m$$

To Find:

- Surface temperature, T.
- Monochromatic Emissive Power, $E_{b\lambda}$.

Formula to be used:

(a) From Wien's law,

$$\lambda_{max} T = 2.9 \times 10^{-3} \text{ m k}$$

$$T = (2.9 \text{ x } 10^{-3})/(0.52 \text{ x } 10^{-6})$$
$$T = 5576 \text{ k}$$

(b) Monochromatic Emissive Power (E_{b2}),

From Planck's law,

$$E_{b\lambda} = (C_1 \lambda^{-5}) / \left(C^{(c_2/\lambda T)} - 1\right)$$

Where,

$$C_1 = 0.374 \times 10^{-15} \text{ W m}^2 \lambda = 0.52 \times 10^{-6} \text{ m}$$
$$C_2 = 14.4 \times 10^{-3} \text{ m kt} = 5576 \text{ K}$$
$$E_{b\lambda} = 0.374 \text{ x } 10^{-15} \text{ x } (0.52 \text{ x } 10^{-6}) \; / \; e^{(14.4 \text{ x } 10-3) / 0.52\text{x } 10-6 \text{ x}(5576)} - 1$$
$$= 6.9 \times 10^{13} \text{ W/m}^2$$

Result:

- T = 5576 K.
- $E_{b2} = 6.9 \times 10^{13} \text{ W/m}^2$.

3. A black body at 3000 K emits radiation. Let us calculate monochromatic emissive Power at 1μm wave length, Wavelength at which emission is Maximum, Maximum Emissive Power, Total Emissive Power and also calculate the total Emissive of the furnace if it is assumed as a Real Surface having Emissivity equal to 0.85.

Solution:

Given:

Surface temperature, T = 3000 K

To Find:

- Monochromatic emissive Power (E_{b2}) at $\lambda = 1; \mu = 1 \times 10^{-6}$ m
- Maximum Wave length (λ_{max})

- Maximum Emissive Power $(E_{b\lambda})_{max}$
- Total Emissive Power (E_b)
- Emissive Power of real surface at $\varepsilon = 0.85$
- Monochromatic Emissive Power

1) From Planck's distribution law, we know that,

$$E_{b\lambda} = \frac{c_1 \lambda^{-5}}{\left[\ell^{\left(\frac{c_2}{\lambda T}\right)}_{-1} \right]}$$

Where,

$$C_1 = 0.374 \times 10^{-15}\ \text{W m}^2$$

$$C_2 = 14.4 \times 10^{-3}\ \text{m K}$$

$$\lambda = 1 \times 10^{-6}\text{M};\ T = 3000\ \text{k}$$

$$E_{b\lambda} = \frac{0.374 \times 10^{-15} \left[1 \times 10^{-6}\right]^{-5}}{\left(\ell^{\left[\frac{14.4 \times 10^{-3}}{1 \times 10^{-6} \times 3000}\right]}_{-1} \right)}$$

$$E_{b\lambda} = 3.10 \times 10^{12}\ \text{W/m}^2$$

2) Maximum Wave Length $(\lambda)_{Max}$

From Wien's law, we know that,

$$\lambda_{max} T = 2.9 \times 10^{-3}\ \text{m K}$$

$$\lambda_{max} = \frac{2.9 \times 10^{-3}}{3000}$$

$$\lambda_{max} = 0.966 \times 10^{-6}\ \text{m}$$

3) Maximum Emissive Power $(E_{b\lambda})_{max}$

$$(E_{b\lambda})_{max} = 1.307 \times 10^{-5} \times T^5$$

$$= 1.307 \times 10^{-5} \times (3000)^5$$

$$(E_{b\lambda})_{max} = 3.17 \times 10^{12}\ \text{W/m}^2$$

4) Total Emissive Power (E_b)

From Stefan Boltzmann law, we know that

$$E_b = \sigma T^4$$

Where,

$$\sigma \rightarrow \text{Stefan Boltzmann Constant} = 5.67 \times 10^{-8} W/m^2K^4$$

$$\therefore E_b = 5.67 \times (10)^{-8} \times (3000)^4$$

$$E_b = 4.59 \times 10^6 \ W/m^2$$

5) Total Emissive Power of a Real Surface

$$(E_b)_{real} = \varepsilon \sigma T^4$$

Where,

$$\varepsilon - \text{Emissivity} - 0.85$$

$$\therefore (E_b)_{real} = 0.85 \times 5.67 \times 10^{-8} \times (3000)^4$$

$$(E \ )_{real} = 3.90 \times 10^6 \ W/m^2$$

Result:

- $E_{b\lambda} = 3.10 \times 10^{12} \ W/m^2$
- $\lambda_{max} = 0.966 \times 10^{-6} m$
- $(E_{b\lambda})_{max} = 3.17 \times 10^{12} W/m^2$
- $E_b = 4.59 \times 10^6 W/m^2$
- $(E_b)_{real} = 3.90 \times 10^6 W/m^2$

4.6 Refrigeration System: Reversed Carnot Cycle

A reversed heat engine cycle is visualized as an engine operating in the reverse way, i.e. receiving heat from a low temperature region, discharging heat to a high temperature region and receiving a net inflow of work shown in the figure. Under such conditions the cycle is called a heat pump cycle or a refrigeration cycle.

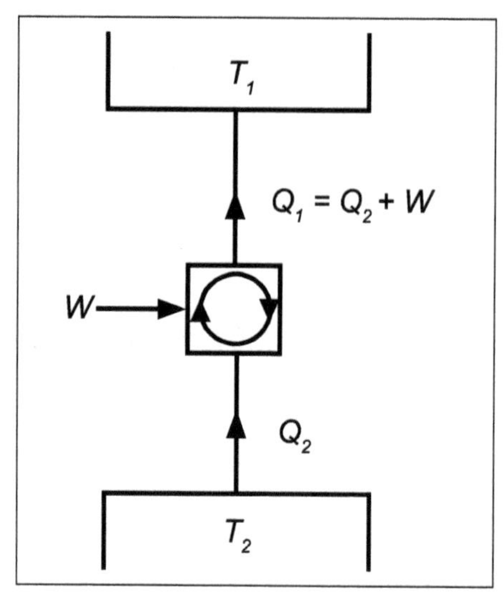

Reversed heat engine cycle.

$$(COP)_{H.P.} = \frac{Q_1}{W} = \frac{Q_1}{Q_1 - Q_2}$$

And for a refrigeration,

$$(COP)_{ref} = \frac{Q_2}{W} = \frac{Q_2}{Q_1 - Q_2}$$

As studied earlier if the direction of Carnot engine is reversed, then the cycle works as a refrigeration cycle. The reversed Carnot cycle will theoretically have a maximum possible coefficient of performance, but it is not possible to construct a refrigerating machine which will work on the reversed Carnot cycle.

A reversed Carnot cycle using air as working medium is shown on p—V and T—S diagrams respectively.

The Cycle Consists of Four Reversible Processes in Sequence

Process 1-2: Isentropic expansion of air from higher temperature T_H to lower temperature T_L.

Process 2-3: Heat removal from cold space in isothermal manner at temperature T_L.

Process 3-4: Isentropic compression of air from low temperature T_L to high temperature T_H.

Process 4-1: Heat rejection isothermally to a medium at temperature T_H.

The refrigeration effect = Heat absorbed by air during isothermal process 2-3 at temperature T_2.

$$RE = T_L(S_3 - S_1)$$

Heat rejected at temperature T_H,

$$QH = T_H(S_3 - S_2)$$

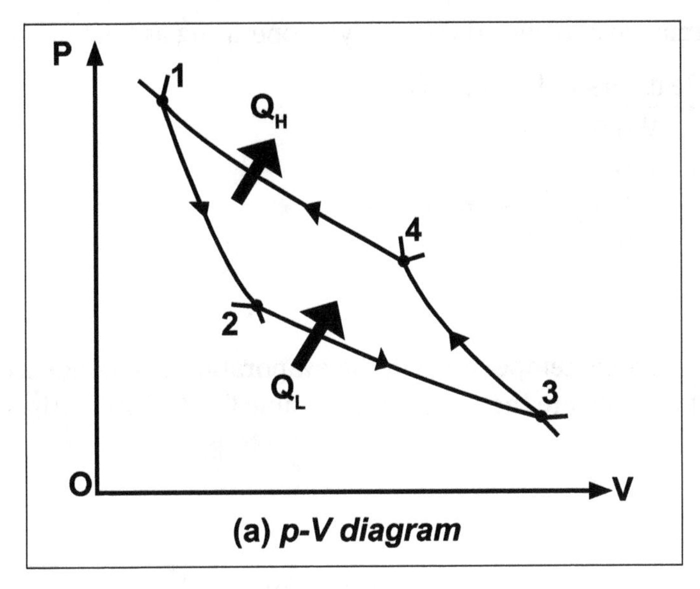

(a) p-V diagram

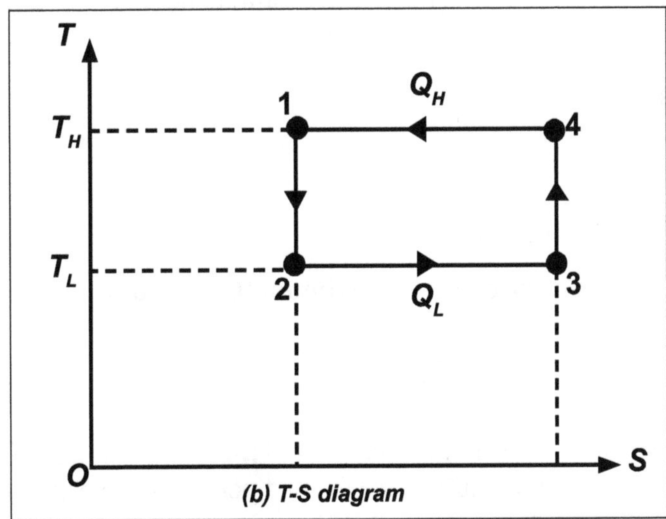

(b) T-S diagram

Reversed Carnot cycle

Net work input to cycle,

$$
\begin{aligned}
W_{in} &= Q_H - Q_L \\
&= T_H(S_3 - S_2) - (S_3 - S_2) \\
&= (T_H - T_L)(S_3 - S_2)
\end{aligned}
$$

The coefficient of performance of reversed Carnot cycle operating as refrigerator is given by,

$$(COP)_{R.rev} = \frac{RE}{W_{in}} = \frac{T_L(S_3 - S_2)}{(T_H - T_L)(S_3 - S_2)}$$

$$= \frac{T_L}{T_H - T_L}$$

The coefficient performance of reversed Carnot cycle operating as heat pump is given by,

$$(COP)_{R.rev} = \frac{\text{Heat rejected at } T_H}{\text{Work input}} = \frac{Q_H}{W_{in}}$$

$$= \frac{T_H(S_3 - S_2)}{(T_H - T_L)(S_3 - S_2)} == \frac{T_H}{T_H - T_L}$$

Problems

1. A refrigerator has a working temperature in the evaporation and condenser coils as - 30°C and 30°C respectively. Let us calculate the maximum possible COP of the refrigerator.

Solution:

Given:

Refrigerator working temperature in the evaporation and condenser coils

$$T_L = -30^{\circ}C = 243 \text{ K}$$
$$T_H = 30^{\circ}C = 303 \text{ K}$$

To find:

Maximum possible COP of a refrigerator.

Analysis: The reversed Carnot cycle can give maximum $(COP)_R$, thus

$$(COP)_{R.rev} = \frac{TL}{T_H - T_L} = \frac{243}{303 - 243} = 4.05$$

2. A reversed Carnot cycle is used for making ice at -5°C from water at 25°C .The temperature of the brine is -10°C. Let us calculate the quantity of ice formed per kWh of work input. Assume the specific heat of ice as 2 kJ/kg.k, latent heat of ice as 335 kJ/Kg and specific heat of water as 4.18 kJ/kgK.

Solution:

Given:

Formation of ice at -5°C from water at 25°C

T_L = -10° C = 263 K

T_H = 25° C = 298 K

T_{ice} = -5° C

W_{in} = 1 KWh =3600kJ

$C_{p.ice}$ = 2 kJ/kg.k

C_{pw} = 4.18 kJ/kg.k

H_{fg} = 335 kJ/kg

To find:

The mass of ice formed.

Assumption: The ambient temperature to be 25°C.

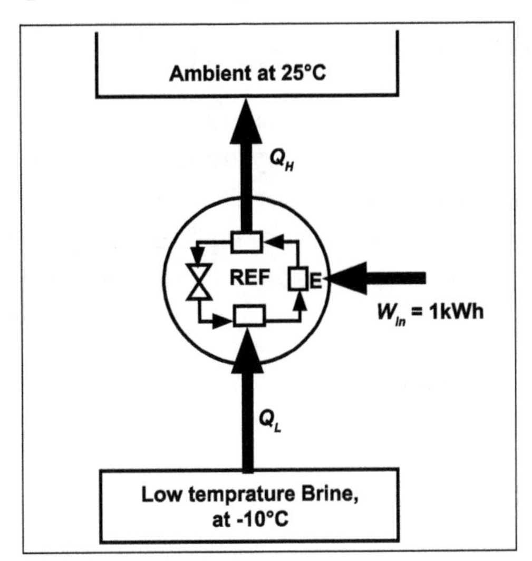

Amount of heat removed from water-ice system per kg.

Q_L = Heat removed in cooling of water from 25°C to 0°C + heat removed from ice during it is cooling from 0°C to -5°C

$$= C_{pw}\left(T_H - 0^\circ C\right) + h_{fg} + C_{p.ice}\left(0 - T_{ice}\right)$$
$$= 4.18 \times (25 - 0) + 355 + 2.0 \times \left[0-(-5)\right]$$
$$= 449.5 \text{ kJ/kg.}$$

The COP of a Carnot refrigerator,

$$(COP)_{R.rev} = \frac{T_L}{T_H - T_L}$$

$$= \frac{263}{298-263}$$

Further, the COP can also be expressed as,

$$COP = \frac{RE}{W_{in}}$$

$$7.514 = \frac{RE}{3600\,kJ}$$

$$RE = 27051.43 \text{ kJ}$$

Thus, the mass of ice formed with this refrigerating effect is given by,

$$\dot{m}_{ice} = \frac{RE}{Q_L} = \frac{27051.43}{449.5} = 60.18 \text{ kg / kWh}$$

4.6.1 Reversed Brayton Cycle (Gas refrigeration System)

The Bell Coleman refrigeration cycle is the reverse of the closed Brayton power cycle. The schematic and T-s diagrams of the reverse Brayton cycle are shown in the figure (a). The refrigerant gas (may be air) enters the compressor at the state 1 and is compressed to the state 2. The gas is then cooled at constant pressure in a heat exchanger to the state 3. During cooling, the gas rejects heat to the surroundings and approaches the temperature of the warm environment.

The gas is then expanded in an expander to the state 4, where it attains a temperature that is well below the temperature of the cold region. The refrigeration effect is achieved through the heat transfer from the cold region to gas as it passes from state 4 to 1 in a heat exchanger and the cycle completes.

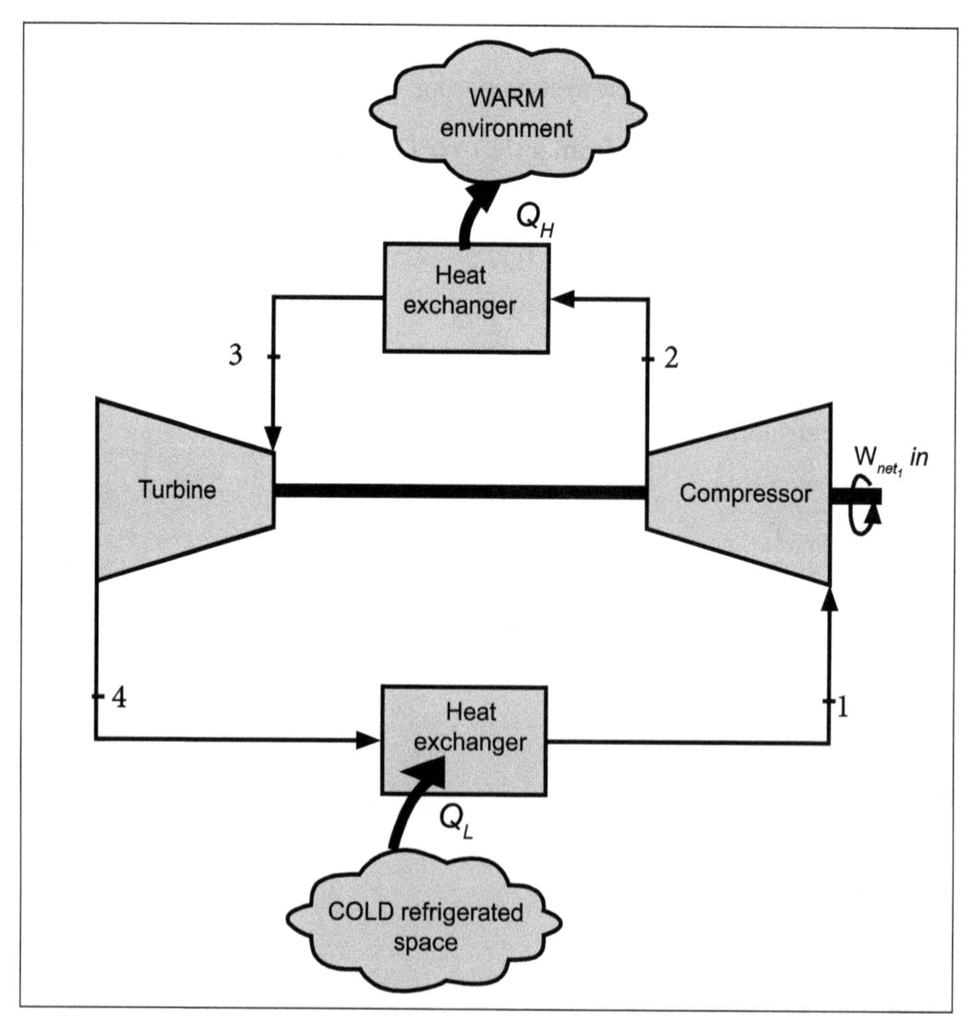

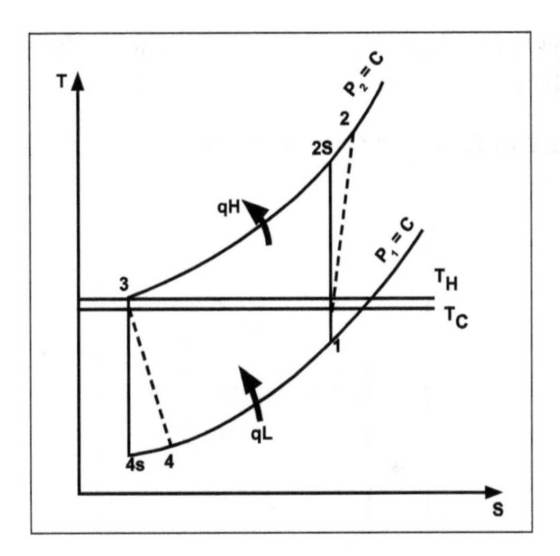

(a) Brayton refrigeration cycle

The T-s diagram for an ideal Brayton cycle is represented by the cycle 1-2s-3-4s-1, in which all processes are internally reversible and compression and expansion are isentropic. The cycle 1-2-3-4-1 includes the effect of irreversibilities during adiabatic compression and expansion. The frictional pressure drops have been ignored.

Cycle Analysis At steady state, the compression and expansion work per kg of gas flow in the system are,

$$w_c = h_2 - h_1 \text{and } w_T = h_3 - h_4$$

The network input the cycle is given by,

$$\begin{aligned} w_{net} &= w_c - w_T \\ &= (h_2 - h_1) - (h_3 - h_4) \end{aligned}$$

The refrigeration effort produced in the cycle,

RE = Heat transfer from the cold region to the refrigerant gas = $h_1 - h_4$

The coefficient of performance of the cycle is the ratio of the refrigeration effect to the net work input,

$$(COP)_R = \frac{RE}{w_{net}} = \frac{h_1 - h_4}{(h_2 - h_1) - (h_3 - h_4)}$$

The gas refrigeration cycle deviates from the re-versed Carnot cycle because heat transfer processes are not isothermal. In fact, the gas temperature varies considerably during the heat-transfer process. The figure (b) shows a T-s diagram which compares the reversed Carnot cycle 1-2'-3-4'-1 and reversed Brayton cycle 1-2-3-4-1. It reveals the following facts:

- The net work (area 1-2'-3-4'-1) required by the reverse Carnot cycle is a fraction of that required by the reverse Brayton cycle.

- The reverse Carrot cycle produces greater refrigeration effect (area under line 4'-1) as compare to reverse Brayton cycle (area under curve 4 -1).
- The mean temperature of heat rejection is much greater and that of heat rejection is much lower in reverse Brayton.

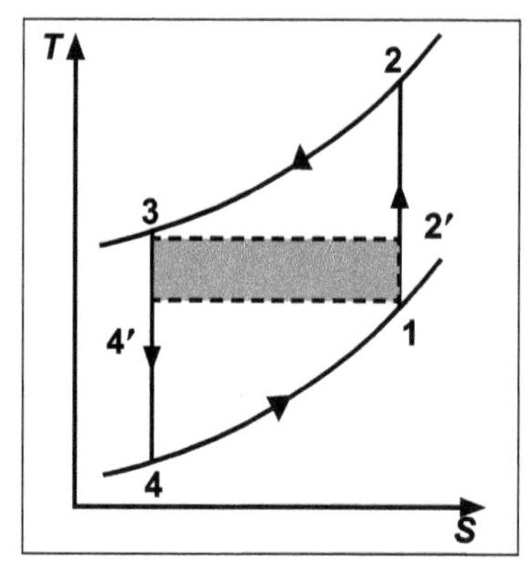

Comparison of revered Brayton cycle and reversed Carnot cycle.

Therefore, the COP of the reverse Brayton cycle is much lower than that of the vapour compression cycle or the reversed Carrot cycle.

Advantages:

- Air is a freely and easily available fluid.
- There is no danger of fire and toxic effects, if it leaks.
- The weight of air refrigeration system per ton of refrigeration is less compared with other refrigeration systems.
- An air cycle can work as an open or closed cycle system.
- It is ecofriendly.

Disadvantages:

- The main drawback of an air refrigeration system is its very low value of COP.
- Air has only sensible heat and cannot transfer heat at constant temperature across heat exchangers. Thus, as the temperature difference decreases, its heat transfer capacity de-creases.
- The quantity of air required per ton of refrigeration capacity is much larger than the liquid refrigerants.
- A turbine is required for the expansion of air instead of a throttling device. Since air contains some water vapour, thus danger of frosting during expansion is more possible.
- It has more running cost than other systems.

Problems

1. A gas turbine works on an air standard Brayton cycle. The initial condition of the air is 25° C and

1 bar. The maximum pressure and temperature are limited to 3 bar and 650° C. Let us determine the following:

- Cycle efficiency.
- Heat supplied and rejected per kg of air.
- Work output.
- Exhaust temperature.

Solution:

Given data:

$p_1 = p_4 = 1$ bar = 100 kN/m²

$T_1 = 25°C = 298$ K

$p_2 = p_3 = 3$ bar = 300 kN/m²

$T_3 = 650°C = 923$ K

To find:

- Cycle efficiency.
- Heat supplied and rejected per kg of air.
- Work output.
- Exhaust temperature.

Consider the Process 1 - 2 Isentropic Compression

$$\frac{T_2}{T_1} = \left(\frac{P_2}{P_1}\right)^{\frac{\gamma-1}{\gamma}}$$

$$T_2 = \left(\frac{P_2}{P_1}\right)^{\frac{\gamma-1}{\gamma}} \text{ x } T_1$$

$$= \left(\frac{300}{100}\right)^{\frac{1.4-1}{1.4}} \text{ x } 298$$

$$T = 407.88\,\text{K}$$

Consider the Process 3 - 4 Isentropic Expansion

$$\frac{T_4}{T_3} = \left(\frac{P_4}{P_3}\right)^{\frac{\gamma-1}{\gamma}}$$

$$T_4 = \left(\frac{P_4}{P_3}\right)^{\frac{\gamma-1}{\gamma}} x\, T_3$$

$$= \left(\frac{100}{300}\right)^{\frac{1.4-1}{1.4}} x\, 923$$

$$T_4 = 1263.34K$$

Work output

$$W_C = C_P(T_2 - T_1)$$

$$= 1.005 \times (1263.34 - 298)$$
$$W_c = 970.16 \text{ kJ}$$

Heat supplied

$$Q_s = C_P(T_4 - T_3)$$
$$= 1.005 \times (1263.34 - 923)$$
$$Q_s = 342.04 \text{ kJ/kg}$$

Heat rejected

$$Q_r = Cp\,(T_6 - T_1)$$
$$= 1.005\,(407.881 - 298)$$
$$Q_R = 110.43 \text{ kJ/kg}$$

4.7 The Vapor Compression Cycle

The vapour Compression Refrigeration System consists of the following five essential parts:

- Compressor.
- Condenser.
- Receiver.
- Expansion Valve.
- Evaporator.

1. Compressor

The low pressure and temperature vapour refrigerant from evaporator is drawn into the compressor through the inlet or suction valve where it is compressed to a high pressure and temperature. In this high pressure and temperature, vapour refrigerant is discharged into the condenser through the delivery or discharge valve B.

2. Condenser

The condenser or cooler consists of coils of pipe in which the high pressure and temperature vapour refrigerant is cooled and condensed. The refrigerant, while passing through the condenser, gives up its latent heat to the surrounding condensing medium which may be normally air or water.

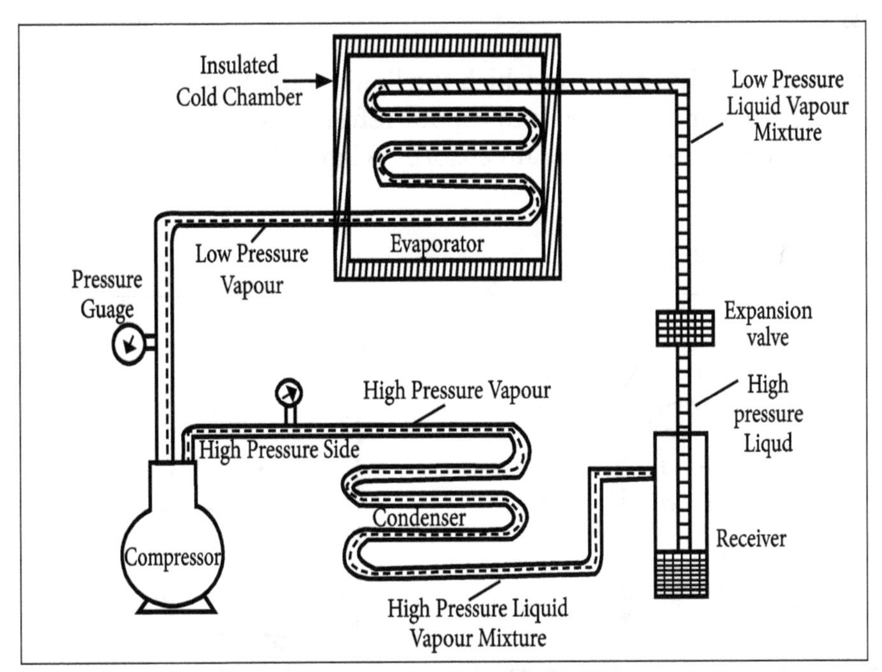

Vapour Compression Refrigeration System.

T-s and p-h diagrams for the vapour compression cycle when the vapour after compression is dry saturated.

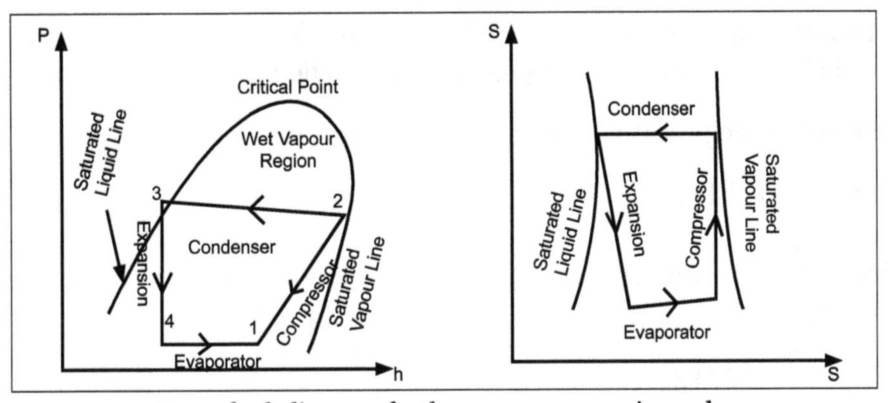

T-s and p-h diagrams for the vapour compression cycle.

3. Receiver

The condensed liquid refrigerant from the condenser is stored in a vessel known as receiver from where it is supplied to the evaporator through the expansion valve or refrigerant control valve.

4. Expansion Valve

It is also called as throttle valve or refrigerant control valve. The function of the expansion valve is to allow the liquid refrigerant under high pressure and temperature to pass at a controlled nature after reducing its pressure and temperature. Some of the liquid refrigerant evaporates as it passes through the expansion valve but the greater portion is vaporized in the evaporator at the low pressure and temperature.

5. Evaporator

An evaporator consists of coils of pipe in which the liquid-vapour refrigerant at low pressure and temperature is evaporated and changed into vapour refrigerant at low pressure and temperature. In evaporating, the liquid vapour refrigerant absorbs is latent heat of vaporization from the medium (air, water or brine) which is to be cooled.

The basic operations for ideal plants are:

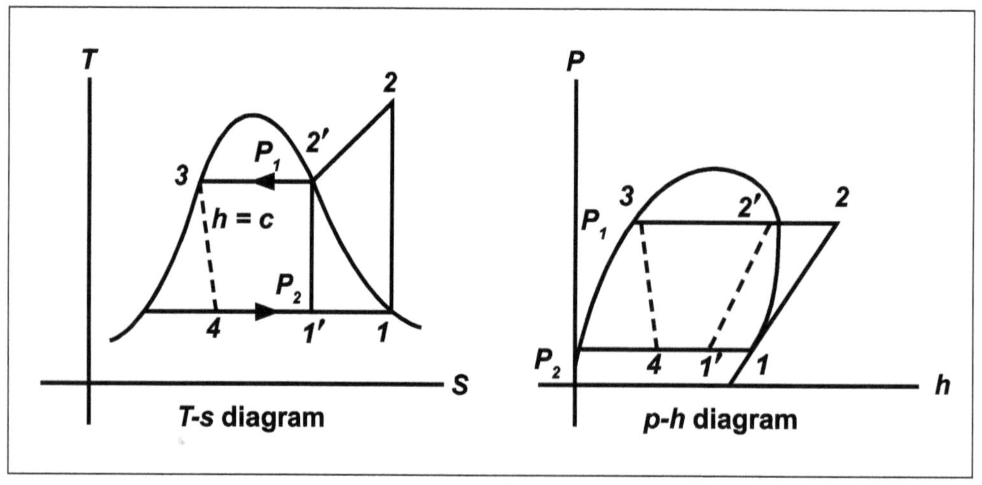

T-s and P-h diagrams.

1. Compression: Process 1-2: The saturated vapour of refrigerant is compressed isentropically. Process 1-2 is called dry compression but process 1'-2' is called wet compression.

The compressor can be reciprocating, rotary or centrifugal type.

$$W_c = (h_2 - h_1)\ [kJ/kg]$$

2. Condensing: Process 2-3: The high pressure refrigerant gas is first de-superheated and condensed. The heat rejected,

$$Q_1 = (h_2 - h_3)\ [kJ/kg]$$

3. Expansion: Process 3-4: The high pressure liquid refrigerant is expanded in an expansion valve or capillary tube. The process of throttling is adiabatic,

$$h_3 - h_4$$

The quality of refrigerant at inlet to evaporator,

$$X_4 = \frac{h_{f3} - h_{f4}}{h_{fg_4}}$$

4. Evaporation: Process 4-1- The evaporation process is a constant pressure process. The heat is picked up from surrounding at high temperature by refrigerant at low temperature after throttling in expansion valve. The heat picked up is called refrigeration effect Q_2.

The liquid refrigerant is evaporated into low pressure gas and enters the compressor to start a new cycle.

$$Q_2 = (h_1 - h_4)\ [kJ/kg]$$

The values of h_1, h_2, h_3 and h_4 can be obtained from the property tables and charts of the refrigerant.

$$COP = \frac{Q_2}{W_c} = \frac{(h_1 - h_4)}{(h_2 - h_1)}$$

Advantages:

- It has less running cost.
- It has smaller size for the given capacity of refrigeration.
- The coefficient of performance is quite high.
- It can be employed over a large range of temperatures.

Disadvantages:

- The prevention of leakage of the refrigerant is the major problem in vapour compression system.
- Its initial cost is high.

4.7.1 The Vapor Absorption Cycle

The idea of a vapour absorption refrigeration system is to avoid compression of the refrigerant. In this type of refrigeration system, the vapour produced by the evaporation of the refrigerant in the cold chamber, passes into a vessel containing a homogeneous mixture of ammonia and water (known as aqua-ammonia). In this chamber, the vapour is absorbed, which maintains constant low pressure, thus facilitating its further vaporization.

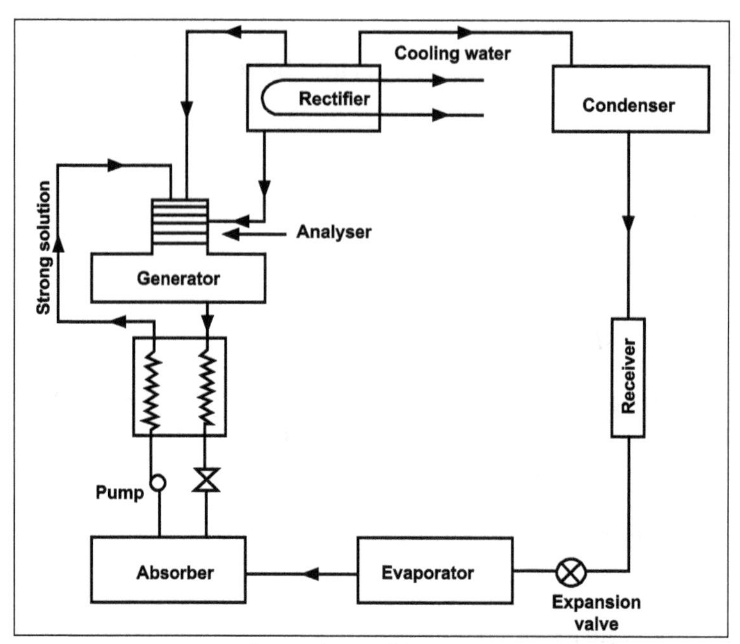

Vapour Absorption Refrigeration System.

The refrigerant is liberated in the vapour state subsequently by the direct application of heat and at such a pressure condensation can be effected at the temperature of the air by cold water. The low pressure ammonia vapour, leaving the evaporator, enters the absorber where it is absorbed in the weak ammonia solution. This process takes place at a temperature slightly above than that of the surroundings. In this process, some heat is transferred to the surroundings.

The strong ammonia solution is then pumped through a heat exchanger to the generator, where a high pressure and temperature is maintained. Under these conditions, the ammonia vapour is driven from the solution. This happens because of the heat transfer from a high temperature source.

The ammonia vapour enters into the condenser, where it gets condensed, in the same way as in the vapour compression system. The weak ammonia solution returns back to the absorber through a heat exchanger.

The equipment used in a vapour absorption system is somewhat complicated than in a vapour compression system. It can be economically justified only in those cases where a suitable source of heat is available which would be wasted.

The coefficient of performance of this refrigerator is given by,

$$\text{COP} = \frac{\text{Heat absorbed during evaporation}}{\text{Work done by pump} + \text{heat supplied in heat exchanger}}$$

Mathematically,

$$\text{COP} = \frac{T_3(T_1 - T_2)}{T_1(T_2 - T_3)}$$

Where,

T_1 = Temperature at which the working substance receives heat.

T_2 = Temperature of the cooling water.

T_3 = Evaporator temperature.

Merits:

- No need of electric power.
- Wear and tear is less.
- Ton capacity is high.
- There is no leakage of refrigerant.
- At part loads, the performance is not affected.
- Space requirement is less.

Demerits:

- Energy requirement is high.
- Charging of refrigerant is difficult.

Comparison between Vapour Compression and Vapour Absorption Systems

Particulars	**Vapour compression system**	**Vapour absorption system**
Type of energy supplied.	Mechanical - a high grade energy.	mainly heat-a low grade energy.
Energy supply	Low	High
Wear and tear	More	Less
Performance at part loads	Poor	System not affected by variations of loads.
Suitability	Used where high grade mechanical energy is available.	Can also be used at remote places as it can work even with a simple kerosene lamp (of course in small capacities).
Charging of refrigerant	Simple	Difficult
Leakage of refrigerant	More chances	No chance as there is no compressor or any reciprocating component to cause leakage.
Damage	Liquid traces in suction line may damage the compressor.	Liquid traces of refrigerant present in piping at the exit of evaporator constitute no danger.

4.7.2 Air Conditioning

Working Principle

An air conditioner became compulsory home appliance and starting from schools, offices to 1 BHK household apartments everywhere we will find these appliances. Main function of any air-conditioner is to control the temperature and humidity of the air that is flowing into the room in which it is placed. People generally use it for temperature control in enclosed room. Popularly known as A.C., this device is popular in almost all the countries. There are several types of A.C.s like split, window, centralized etc. based on the design and capacity. The principle behind working of A.C. is same as that of refrigerator.

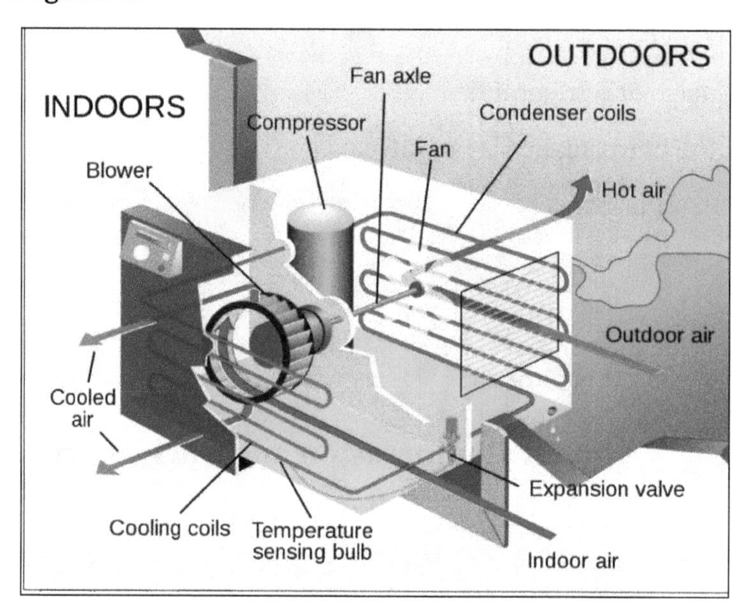

Main Parts Of the Equipment in the Air Conditioning Cycle

1. Circulating Fan

The main function of this fan is to remove air to and from the room.

2. Air Conditioning Unit

It is a unit, which consists of cooling and de-humidifying processes for summer air conditioning or heating and dehumidification processes for winter air conditioning.

3. Supply Duct

It directs the conditioned air from the circulating fan to the space to be air-conditioned at proper point.

4. Supply Outlets

These are grills, which distribute the conditioned air evenly in the room.

5. Return Outlets

These are the openings in a room surface which allow the room air to enter the return duct.

6. Filters

The main function of the filters is to remove dust, dirt and other harmful bacteria's from the air.

Summer Air Conditioning System

Summer Air conditioning system is the most important type of air conditioning in which the air is cooled and generally dehumidified. The schematic arrangement of a typical summer air conditioning is shown in figure.

This system consists of air dampers, filters, cooling coil and dehumidifier, Heating coil and fans. The outside air flows through the damper and mixes up with the recirculated air (which is obtained from the conditioned space). The mixed air passes through a filter to remove dirt, dust and other impurities. The air now passes through a cooling coil. The coil has a temperature much below that required dry bulb temperature of the air in the conditioned space.

The cooled air passes through a perforated membrane and losses its moisture in the condensed form which is collected in a sump. After that, the air is made to pass through a heating coil which heats up the air slightly. This is done to bring the air to the designed dry bulb temperature and relative humidity.

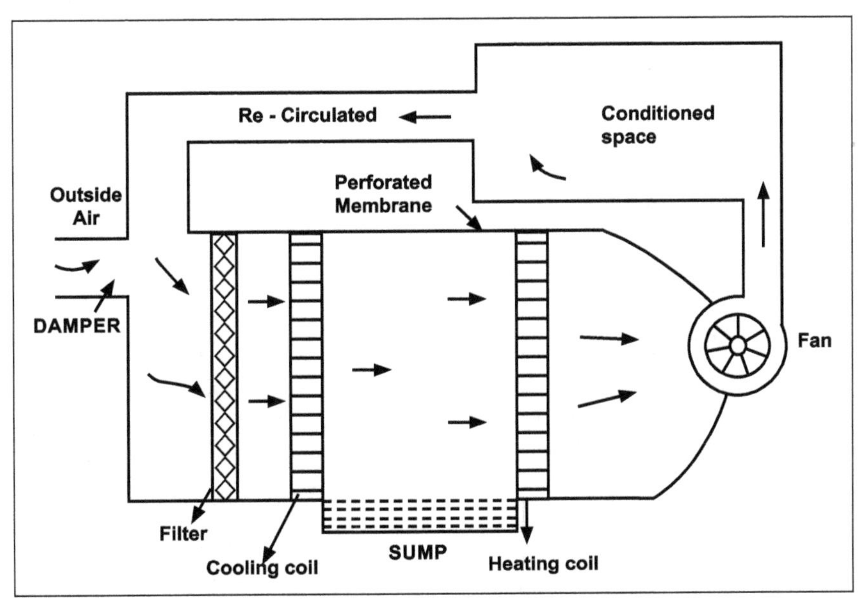

Summer Air Conditioning System.

Now, the conditioned air is supplied to the conditioned space by a fan. From the conditioned space, a part of the used air is exhausted to the atmosphere by the exhaust fan or ventilators. The remaining part of the used air (known as recirculated air) is again conditioned. The outside air is sucked and made to mix with the recirculated air in order to make up for the loss of conditioned (or used) air through exhaust fans or ventilation from the conditioned space.

Winter Air Conditioning System

In winter air conditioning, the air is heated, which is generally accompanied by humidification. The schematic arrangement is shown in the figure below.

The outside air flows through a damper and mixes up with the recirculated air (which is obtained from the conditioned space). The mixed air passes through a filter to remove dirt, dust and other impurities. The air now passes through a pre-heat coil in order to prevent the possible freezing of water and to control the evaporation of water in the humidifier.

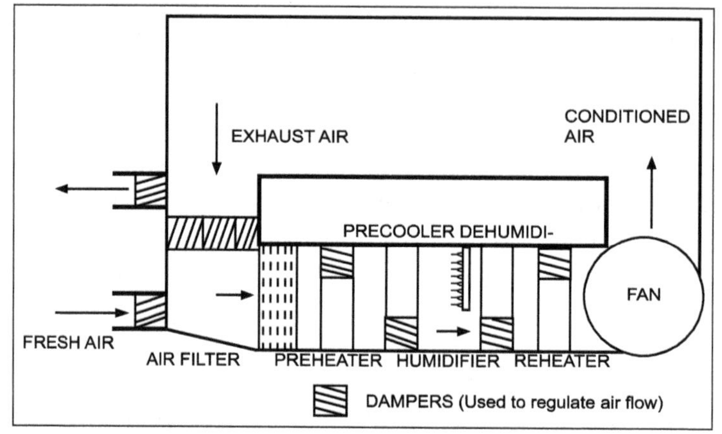

Winter Air Conditioning System.

After that, the air is made to pass through a reheat coil to bring the air to the designed dry bulb temperature. Now the conditioned air is supplied to the conditioned space by a fan. From the conditioned space, a part of the used air is exhausted to the atmosphere by the exhaust fans or ventilators. The remaining part of the used air (known as recirculated air) is again conditioned.

The outside air is sucked and made to mix with the recirculated air, in order to make up for the loss of conditioned (or used) air through exhaust fans or ventilation from the conditioned space.

Window Type Air Conditioner

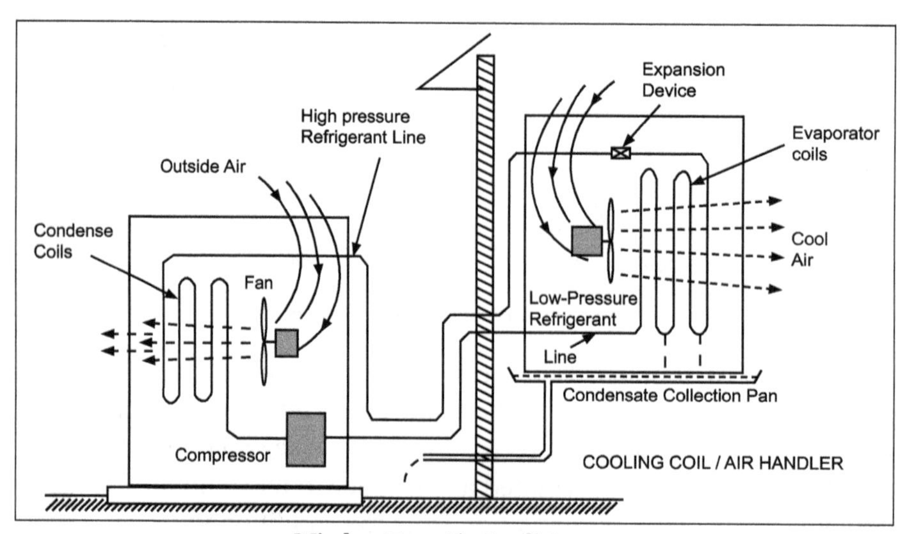

Window Type Air Conditioner.

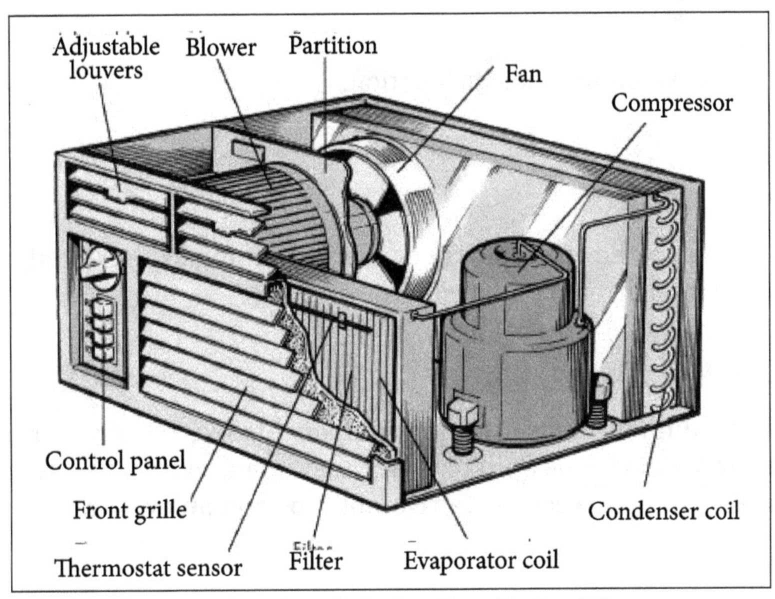

Cross section view of Window Type Air Conditioner.

Construction

This is also called room air conditioner. This unit consists of a cooling system to cool and dehumidify the air involves a condenser, a compressor and a refrigerant coil. A filter is used to remove any impurities in the air. The filter is made of mesh, glass wool or fiber. A fan and adjustable grills are used to circulate the air. Controls are used to regulate the equipment operation.

The low pressure refrigerant vapour is drawn from the evaporator to the hermetic compressor through suction pipe. It is compressed from low pressure to the high pressure and supplied to the condenser. It is condensed in the condenser by passing the outdoor air over the condenser coil by a fan.

The liquid refrigerant is passed through the capillary into the evaporator. In the evaporator, the liquid refrigerant picks up the heat from the refrigerator surface and gets vaporized. A motor driven fan draws air from the room through the air filter and this air is cooled by losing its heat to the low temperature refrigerant and cold air is circulated back into the room.

The vapour refrigerant from the evaporator goes to the compressor from evaporator and the cycle is repeated. Thus, the room is air conditioned. The quantity of air circulated can be controlled by the dampers. The moisture in the air passing over the evaporator coil is dehumidified and drips into the trays. This evaporates water to certain extent and thus, helps in cooling the compressor and condenser.

The unit automatically stops when the required temperature is reached in the room. This is accomplished by the thermostat and control panel.

Merits:

- A separate temperature control is provided in each room.
- Ducts are not required for distribution.

- Cost is less.
- Skilled technician is required for installation.

Demerits:

- It makes noise.
- Large hole is made in the external wall or a large opening to be created in the window panel. This leads to insecurity to inmates.

Split Type Air Conditioner

In split air type air conditioner, noise making components like compressor and condenser are mounted outside or away from room. Split type air conditioning system has two main components. The indoor unit consists of power cables, refrigerant tube and an evaporator mounted inside the room.

Working

Compressor is used to compress the refrigerant. The refrigerant moves between the evaporator and condenser through the circuit of tubing and fins in the coils. The evaporator and condenser are usually made of coil of copper tubes and surrounded by aluminum fins. The liquid refrigerant coming from the condenser evaporates in the indoor evaporator coil. During this process, the heat is removed from the indoor unit air and thus, the room is cooled.

Air return grid takes in the indoor air. Water dehumidified out of air is drained through the drain pipe. The hot refrigerant vapour is passed to the compressor and then to the condenser where it becomes liquid. Thus, the cycle is repeated. A thermostat is used to keep the room at a constant, comfortable temperature avoiding the frequent turning on off.

Merits:

- It is compact.
- It saves energy and money.
- Duct is not used.
- Easier to install.
- It is noiseless because rotary air compressor used is kept outside.
- It is more efficient and powerful.
- It has the flexibility for zoning.

Demerits:

- Initial cost is higher than window air conditioner.
- Skilled technician is required for installation.
- Each zone or room requires thermostat to control the air cooling.

Applications of Air Conditioning

- Used in houses, hospitals, offices, computer centers, theaters, departmental stores etc.,
- Air-conditioning of transport media such as buses, cars trains, aero-planes and ships.
- Wide application in food processing, printing, chemical, pharmaceutical and machine tool, etc.

Problems

1. For a summer air conditioning installation for industrial application the following data is given:

Room Design	50% RH 26°C DBT
Outside Design	40°C DBT 10% RH
Room sensible heat gain	40 kW
Room latent heat loss	10 kW

50% of return air from the room is mixed with outdoor air and pre-cooled sensibly in a cooling coil to 28°C before being passed through adiabatic washer. Let us determine:

- Supply air conditions to the space.
- Quantity of fresh outside air.
- Refrigerating capacity of the pre-cooler coil.
- Humidifying efficiency of the adiabatic washer or evaporator cooler and entering and leaving conditions at the washer.

Solution:

Given data:

$td_2 = 26°\ C\ DBT$

$\phi_2 = 50\ \%$

$td_1 = 40\ °\ C\ DBT$

$\phi = 10\ \%$

$R_{SH} = 40\ kW$

$R_{LH} = 10\ kW$

$t\omega_1 = 28°\ C$

To find:

- Supply air conditions to the space.
- Quantity of fresh outside air.
- Refrigerating capacity of the preorder coil.
- Humidifying efficiency.

The flow diagram for the processes involved in the air conditioning of a room.

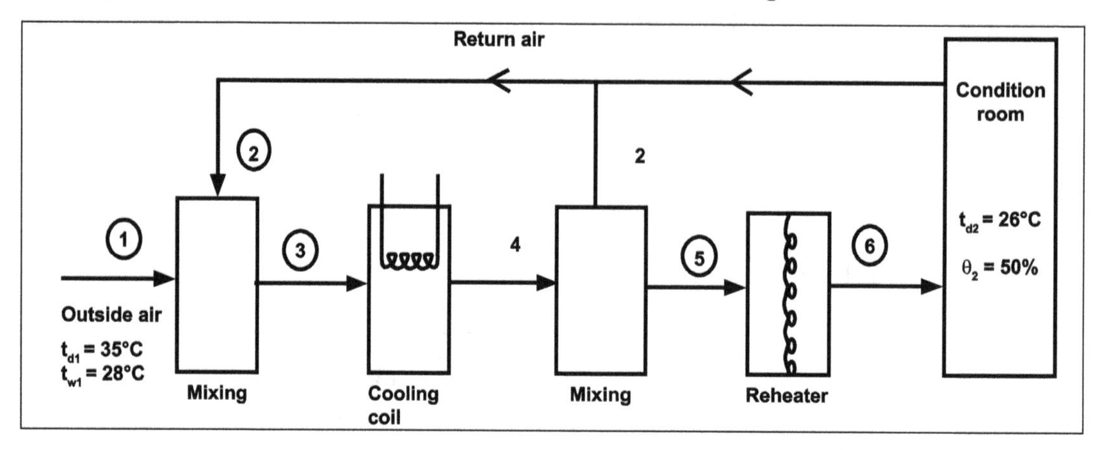

First of all, mark the condition of outside air at 40°C DBT and 28°C WBT on the Psychometric chart at point 1.

Now mark the condition of air inside the room at 26°C DBT and 50% RH as point 2.

Since the return air from the room and the outside air is mixed before extending the cooling coil as 50%, therefore mark the condition of air after mixing and before entering to the cooling coil, (i.e., Point 3) on the line 1-2, such that,

Length 2 – 3 = Length 2 – 1 × 0.5,

$$R_{SHF} = \frac{RSH}{RSH + RLH}$$

$$= \frac{40}{40 + 10} = 0.8$$

$$R_{SHF} = 0.8$$

2. Let us consider a hall to be air-conditioned, the following conditions are given:

Outdoor condition – 40°C dbt, 20°C wbt

Required comfort conditions – 20°C dbt, 60% RH

Seating capacity of hall – 1500

Amount of outdoor air supply – 0.3 m^3 /min/person

If the required condition is achieved first by adiabatic humidification and then by cooling, let us Estimate (i) The capacity of the cooling coil in tonnes and (ii) the capacity of the humidifier is kg/h.

Solution:

Given:

Outdoor Conditions: t_{d1} = 40°C; t_{w1} = 20°C

Required Indoor Conditions: t_{d2} = 20°C; ϕ_2 = 50%.

Amount a free air circulated = 0.3 m³ /min/person

Seating capacity =1500

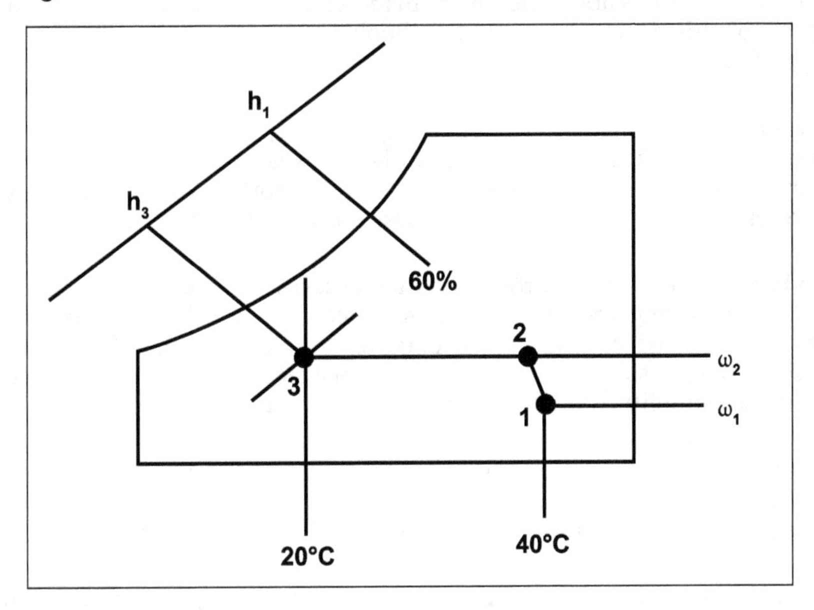

$$6 = \frac{1500 \times 0.3}{0.896} = 502 \text{ kg / min}$$

∴ Capacity of the cooling coil = $6(h_2 - h_3)$

$$= 502\,(57 - 42) \text{ kJ / min}$$

$$= \frac{502 \times 15 \times 60}{14000} = 32.27 \text{ tonnes}$$

∴ Capacity of humidifier:

$= 6(W_2 - W_1)$

$= 502(0.0088 - 0.0065)$ kg/min

$= 502(23 \times 10^{-4}) \times 60$

$= 69.3$ kg/min

Permissions

All chapters in this book are published with permission under the Creative Commons Attribution Share Alike License or equivalent. Every chapter published in this book has been scrutinized by our experts. Their significance has been extensively debated. The topics covered herein carry significant information for a comprehensive understanding. They may even be implemented as practical applications or may be referred to as a beginning point for further studies.

We would like to thank the editorial team for lending their expertise to make the book truly unique. They have played a crucial role in the development of this book. Without their invaluable contributions this book wouldn't have been possible. They have made vital efforts to compile up to date information on the varied aspects of this subject to make this book a valuable addition to the collection of many professionals and students.

This book was conceptualized with the vision of imparting up-to-date and integrated information in this field. To ensure the same, a matchless editorial board was set up. Every individual on the board went through rigorous rounds of assessment to prove their worth. After which they invested a large part of their time researching and compiling the most relevant data for our readers.

The editorial board has been involved in producing this book since its inception. They have spent rigorous hours researching and exploring the diverse topics which have resulted in the successful publishing of this book. They have passed on their knowledge of decades through this book. To expedite this challenging task, the publisher supported the team at every step. A small team of assistant editors was also appointed to further simplify the editing procedure and attain best results for the readers.

Apart from the editorial board, the designing team has also invested a significant amount of their time in understanding the subject and creating the most relevant covers. They scrutinized every image to scout for the most suitable representation of the subject and create an appropriate cover for the book.

The publishing team has been an ardent support to the editorial, designing and production team. Their endless efforts to recruit the best for this project, has resulted in the accomplishment of this book. They are a veteran in the field of academics and their pool of knowledge is as vast as their experience in printing. Their expertise and guidance has proved useful at every step. Their uncompromising quality standards have made this book an exceptional effort. Their encouragement from time to time has been an inspiration for everyone.

The publisher and the editorial board hope that this book will prove to be a valuable piece of knowledge for students, practitioners and scholars across the globe.

Index

Printed in the USA
CPSIA information can be obtained
at www.ICGtesting.com
JSHW051355091023
49903JS00006B/151